NUCLEAR MATERIALS AND DISASTER RESEARCH

PLUTONIUM DISPOSITION

MANAGEMENT, POLICY, AND COST ISSUES

NUCLEAR MATERIALS AND DISASTER RESEARCH

Additional books in this series can be found on Nova's website under the Series tab.

Additional e-books in this series can be found on Nova's website under the e-book tab.

NUCLEAR MATERIALS AND DISASTER RESEARCH

PLUTONIUM DISPOSITION

MANAGEMENT, POLICY, AND COST ISSUES

GLENN REISS
EDITOR

NOVINKA

New York

For permission to use material from this book please contact us:
Telephone 631-231-7269; Fax 631-231-8175
Web Site: http://www.novapublishers.com

Library of Congress Cataloging-in-Publication Data

ISBN: 978-1-63321-066-0

Published by Nova Science Publishers, Inc. † New York

CONTENTS

PREFACE

With the end of the Cold War and breakup of the Soviet Union in the early 1990s, control of surplus nuclear weapons material became an urgent U.S. foreign policy goal. Particular U.S. concern focused on plutonium from Soviet nuclear warheads, which it was feared posed a major nuclear weapons proliferation risk. The United States supported a successful effort to consolidate the storage of Soviet nuclear weapons and materials in Russia, and then began negotiating reductions in weapons material stockpiles. Congress has been closely involved in formulating U.S. policy on surplus plutonium disposition, as well as funding the necessary facilities, operations, and Russian assistance to implement the program. Congressional debate is now focusing on the U.S. program's escalating costs and the Obama Administration's FY2015 proposal to halt construction of plutonium disposition facilities in South Carolina and prepare a new strategy. This book discusses the mixed-oxide fuel fabrication plant and plutonium disposition. It also examines drivers the National Nuclear Security Administration (NNSA) identified for the cost increases of construction projects for plutonium disposition; the extent to which NNSA analyzed underlying causes of the cost increases; steps NNSA took to hold construction contractors accountable for their role, if any, in the cost increases; and the extent to which NNSA's most recent estimates met cost- and schedule-estimating best practices.

Chapter 1 – The Mixed Oxide Fuel Fabrication Facility (MFFF) in South Carolina has been a key component of the current U.S. strategy for disposing of surplus weapons plutonium from the Cold War. That strategy called for the surplus plutonium, in oxide form, to be blended with uranium oxide to make mixed oxide (MOX) fuel for U.S. commercial nuclear reactors. The plutonium in MOX fuel would be mostly destroyed in the reactors by fission (splitting into other isotopes). At the same time, isotopes of plutonium undesirable for weapons would be created, along with highly radioactive fission products. As a result, after several years in a reactor, spent MOX fuel would have less total plutonium than when it was freshly loaded, and the remaining plutonium would be degraded for weapons purposes. Moreover, the fission products would make the material difficult to handle, in case of future attempts to use the plutonium.

Because of sharply rising cost estimates for the MOX project, the Obama Administration is proposing to place MFFF in "cold standby" and study other plutonium disposition options. The federal plutonium disposition program is run by the National Nuclear Security Administration (NNSA), a semiautonomous agency of the Department of Energy (DOE). NNSA estimated in 2002 that MFFF would cost about $1 billion to design and build. DOE said in its budget justification for FY2014 that the MFFF contractor had estimated the project's total construction cost would rise to $7.78 billion, and that construction would not be completed until November 2019. DOE's FY2015 budget justification said the life-cycle cost estimate for the MOX program had risen to $30 billion.

Disposition of surplus plutonium is required by a 1998 agreement, amended in 2010, between the United States and Russia. Each country agreed to convert 34 metric tons of surplus weapons-grade plutonium to a form that could not be returned to nuclear weapons. The two countries agreed to begin plutonium disposition in 2018. The United States is to pay a maximum of $400 million to support the Russian plutonium disposition effort. The Russian government is to fund the remainder of Russia's approximately $2.5 billion in estimated costs. The United States and Russia are now negotiating a verification agreement and a "Milestones" plan that would outline how the U.S. contribution of $400 million to the Russian program would be spent.

Differing sharply from the U.S. MOX strategy, Russia is planning to use its BN-600 and BN-800 fast breeder reactors for plutonium disposition. According to the World Nuclear Association, the BN-800 construction is expected to "start up in 2014 and be operational in 2015," and the reactor is "capable of burning 1.7 metric tons of plutonium per year from dismantled weapons." The DOE FY2015 budget includes no funds for support of the Russian plutonium disposition program. As in FY2014, NNSA plans to use prior-year funds to support these activities in FY2015. The $400 million in U.S. assistance is to be paid to Russia after FY2018 when plutonium disposition is to begin.

The Administration's review of the U.S. plutonium disposition strategy raises several issues for Congress. A fundamental question is whether the rising cost estimates for MFFF are sustainable in the current budget environment. And if the MFFF is not sustainable, what should replace it? However, any delay or major change to the program could affect the planned disposition of Russian weapons plutonium. The effects of budget cuts on DOE's Savannah River Site in South Carolina, where MFFF is located, will also be an important element of the debate.

Chapter 2 – NNSA, a separately organized agency within DOE, manages the Plutonium Disposition program to dispose of surplus weapons-grade plutonium by burning it as MOX fuel—a mixture of plutonium and uranium oxides—in specially modified commercial nuclear reactors. In 2012, DOE forecasted cost increases of close to $3 billion over the previous estimates for the program's two construction projects, the MOX facility and the WSB for disposing of waste from the MOX facility.

GAO was asked to review these cost increases and the life-cycle cost estimate. This report examines: (1) drivers NNSA identified for the cost increases; (2) the extent to which NNSA analyzed underlying causes of the cost increases; (3) steps NNSA took to hold construction contractors accountable for their role, if any, in the cost increases; and (4) the extent to which NNSA's most recent estimates met cost- and schedule-estimating best practices. GAO reviewed NNSA's draft life-cycle cost estimate and contractor estimates of the MOX project's cost and WSB schedule, compared the estimates with cost- and schedule-estimating best practices, and interviewed DOE and NNSA officials.

In: Plutonium Disposition
Editor: Glenn Reiss

ISBN: 978-1-63321-066-0

Chapter 1

MIXED-OXIDE FUEL FABRICATION PLANT AND PLUTONIUM DISPOSITION: MANAGEMENT AND POLICY ISSUES*

Mark Holt and Mary Beth D. Nikitin

SUMMARY

The Mixed Oxide Fuel Fabrication Facility (MFFF) in South Carolina has been a key component of the current U.S. strategy for disposing of surplus weapons plutonium from the Cold War. That strategy called for the surplus plutonium, in oxide form, to be blended with uranium oxide to make mixed oxide (MOX) fuel for U.S. commercial nuclear reactors. The plutonium in MOX fuel would be mostly destroyed in the reactors by fission (splitting into other isotopes). At the same time, isotopes of plutonium undesirable for weapons would be created, along with highly radioactive fission products. As a result, after several years in a reactor, spent MOX fuel would have less total plutonium than when it was freshly loaded, and the remaining plutonium would be degraded for weapons purposes. Moreover, the fission products would make the material difficult to handle, in case of future attempts to use the plutonium.

---

* This is an edited, reformatted and augmented version of a Congressional Research Service publication, CRS Report for Congress R43125, prepared for Members and Committees of Congress, from www.crs.gov, dated March 28, 2014.

Because of sharply rising cost estimates for the MOX project, the Obama Administration is proposing to place MFFF in "cold standby" and study other plutonium disposition options. The federal plutonium disposition program is run by the National Nuclear Security Administration (NNSA), a semiautonomous agency of the Department of Energy (DOE). NNSA estimated in 2002 that MFFF would cost about $1 billion to design and build. DOE said in its budget justification for FY2014 that the MFFF contractor had estimated the project's total construction cost would rise to $7.78 billion, and that construction would not be completed until November 2019. DOE's FY2015 budget justification said the life-cycle cost estimate for the MOX program had risen to $30 billion.

Disposition of surplus plutonium is required by a 1998 agreement, amended in 2010, between the United States and Russia. Each country agreed to convert 34 metric tons of surplus weapons-grade plutonium to a form that could not be returned to nuclear weapons. The two countries agreed to begin plutonium disposition in 2018. The United States is to pay a maximum of $400 million to support the Russian plutonium disposition effort. The Russian government is to fund the remainder of Russia's approximately $2.5 billion in estimated costs. The United States and Russia are now negotiating a verification agreement and a "Milestones" plan that would outline how the U.S. contribution of $400 million to the Russian program would be spent.

Differing sharply from the U.S. MOX strategy, Russia is planning to use its BN-600 and BN-800 fast breeder reactors for plutonium disposition. According to the World Nuclear Association, the BN-800 construction is expected to "start up in 2014 and be operational in 2015," and the reactor is "capable of burning 1.7 metric tons of plutonium per year from dismantled weapons." The DOE FY2015 budget includes no funds for support of the Russian plutonium disposition program. As in FY2014, NNSA plans to use prior-year funds to support these activities in FY2015. The $400 million in U.S. assistance is to be paid to Russia after FY2018 when plutonium disposition is to begin.

The Administration's review of the U.S. plutonium disposition strategy raises several issues for Congress. A fundamental question is whether the rising cost estimates for MFFF are sustainable in the current budget environment. And if the MFFF is not sustainable, what should replace it? However, any delay or major change to the program could affect the planned disposition of Russian weapons plutonium. The effects of budget cuts on DOE's Savannah River Site in South Carolina, where MFFF is located, will also be an important element of the debate.

INTRODUCTION

With the end of the Cold War and breakup of the Soviet Union in the early 1990s, control of surplus nuclear weapons material became an urgent U.S. foreign policy goal. Particular U.S. concern focused on plutonium from Soviet nuclear warheads, which it was feared posed a major nuclear weapons proliferation risk. The United States supported a successful effort to consolidate the storage of Soviet nuclear weapons and materials in Russia, and then began negotiating reductions in weapons material stockpiles.

Congress has been closely involved in formulating U.S. policy on surplus plutonium disposition, as well as funding the necessary facilities, operations, and Russian assistance to implement the program. Congressional debate is now focusing on the U.S. program's escalating costs and the Obama Administration's FY2015 proposal to halt construction of plutonium disposition facilities in South Carolina and prepare a new strategy.

In September 1998, the United States and Russia each agreed to convert 34 metric tons of surplus weapons-grade plutonium to a form that could not be returned to nuclear weapons under the Plutonium Management and Disposition Agreement (PMDA). The agreement was designed to ease concerns about the possible theft or diversion of weapons-grade plutonium by nations or others seeking to develop their own nuclear weapons. According to the agreement, the parties could use two methods for disposing of the plutonium—they could either convert it to mixed oxide (MOX) fuel for nuclear power reactors or immobilize it and dispose of it in a way that would preclude its use in nuclear weapons. However, Russia has expressed little interest in the permanent immobilization and disposal of plutonium, or burning it in conventional reactors, preferring to use the material as fuel for its civilian fast reactor program. The United States initially intended to pursue both immobilization and MOX fuel for conventional reactors. However, after reviewing U.S. nonproliferation policies in 2001, the Bush Administration concluded that the dual approach would be too costly. Instead, it outlined a plan for the United States to convert almost all its surplus plutonium to MOX fuel.

The PMDA was amended with a Protocol which entered into force on July 13, 2011. The PMDA Protocol renews the commitment of each side to dispose of 34 metric tons of weapons-grade plutonium. It also lays out conditions for Russian use of the plutonium as fast reactor fuel, including restrictions on breeding additional plutonium in fast reactors.[1] The two countries agreed to begin plutonium disposition in 2018. The United States is to pay $400 million to support the Russian plutonium disposition. The Russian government is to fund the remainder of the approximately $2.5 billion in estimated costs. The plutonium disposition program, including U.S. assistance for the Russian program, is administered by the National Nuclear Security Administration (NNSA), a semiautonomous agency of the Department of Energy (DOE).

A key component of the U.S. strategy for disposing of surplus weapons plutonium has been the Mixed Oxide Fuel Fabrication Facility (MFFF) at DOE's Savannah River Site in South Carolina. NNSA estimated in 2002 that MFFF would cost about $1 billion to design and build. That estimate had risen to $4.8 billion when the construction of MFFF began in 2007. According to DOE's FY2014 budget request, the MFFF contractor then estimated that the project's total construction cost would rise to $7.78 billion, and that construction would not be completed until November 2019. DOE has slowed construction during FY2013 and FY2014 while other disposition options are evaluated.

DOE's FY2015 budget justification says the total lifecycle cost of the MOX fuel program for plutonium disposition—including construction, operations, waste management, and facility decommissioning—has risen to $30 billion. Because of that increase, according to DOE, "the MOX fuel approach is not viable within the available resources." DOE plans to put MFFF and related facilities currently under construction in "cold standby," reducing funding for the U.S. plutonium disposition program from $560.3 million in FY2014 to $286 million in FY2015. DOE plans to "further analyze options to complete the plutonium disposition mission more efficiently."[2]

RUSSIAN PROGRAM

The Russian Surplus Fissile Materials Disposition program, run by NNSA, is to provide $400 million to help Russia dispose of its surplus weapons plutonium.[3] The DOE FY2015 congressional budget request says that putting the U.S. facility in cold standby "does not diminish" the commitment to the U.S.-Russia Plutonium Management and Disposition Agreement (PMDA), and that the United States will continue to work with Russia to achieve the program's primary goal of Russian disposition of equal quantities of plutonium.

The United States and Russia, along with the International Atomic Energy Agency (IAEA), are negotiating an agreement on verification of plutonium disposition in each country. IAEA will have a role in verifying the quantities of weapons-grade plutonium converted to fuel in both countries. The United States and Russia are also negotiating a "Milestones" plan, which would outline how the U.S. assistance to the Russian plutonium disposition effort would be spent. Negotiations related to the plutonium disposition agreement are led by the Department of State's Bureau of International Security and Nonproliferation together with NNSA. According to DOE's FY2015 budget request, Russia has made "significant progress" toward its PMDA commitments.[4]

Russia is planning to use its BN-600 and BN-800 fast neutron reactors for plutonium disposition. According to the World Nuclear Association, the BN-800 construction is "well-advanced," and the reactor is "capable of burning 1.7 metric tons of plutonium per year from dismantled weapons."[5]

As shown in *Table 1*, DOE requested $10.174 million for the Russian plutonium disposition program for FY2012. The Senate Appropriations Committee recommended $1 million for FY2012, and that level was enacted. The Senate Committee included this reasoning in its report (S.Rept. 112-75):

> *Russian Surplus Materials Disposition.*—The Committee recommends $1,000,000, a reduction of $9,174,000. No funding shall be used to support research and development of the Gas Turbine-Modular Helium Reactor in Russia. The Committee understands that the United States committed $400,000,000, subject to future appropriations, to help Russia dispose of 34 metric tons of excess weapon-grade plutonium, but the Committee will not provide funding for this effort until NNSA can explain how the United States would spend the $400,000,000 and the milestones that Russia must meet before the United States releases any of those funds.

The FY2013 budget justification said that the requested funds would be used to "provide technical support to the DOE in meeting U.S. obligations to support disposition of weapon-grade plutonium in Russia; provide U.S. technical oversight of work in Russia associated with the disposition of surplus Russian weapon-grade plutonium in the BN-600 and BN-800 fast reactors; and support the implementation of IAEA verification activities in both the U.S. and Russia."

The FY2014 and FY2015 budget requests included no funds for the Russian plutonium disposition program. NNSA plans to use prior year funds to support these activities in FY2015. The $400 million in U.S. assistance is to be paid to Russia after FY2018, when plutonium disposition is to begin.

The Russian Federation is to fund the remaining costs of $2 billion, including through contributions from other international donors under the G8 Global Partnership.

Table 1. Funding for the NNSA Russian Plutonium Disposition Program (in $ thousands)

	FY2012	FY2013	FY2014	FY2015 Request	FY2016-FY2019 (projected)
DOE Request	10,174	3,788	0	0	30,000
Senate Appropriation	1,000	3,788	n/a	n/a	n/a
House Appropriation	10,174	0	n/a	n/a	n/a
Enacted	1,000	922	n/a	n/a	n/a

Source: Department of Energy Congressional Budget Justifications; S.Rept. 112-164, H.Rept. 112-462.

Notes: All of the above funds are to be spent in the United States. Appropriations for FY2013, FY2014, and FY2015 are to be supported from prior-year uncosted balances.

U.S. Program

The U.S. Plutonium Disposition Program is also conducted by NNSA. The program had long planned to blend surplus plutonium from the U.S. nuclear weapons program with uranium to make MOX fuel for commercial nuclear reactors.

After fueling a reactor for several years, the plutonium in MOX fuel would be mostly destroyed by fission, and remaining plutonium would be isotopically undesirable for weapons use. In addition, the used MOX fuel would contain highly radioactive fission products that would make it difficult to divert for weapons purposes.

Initially, the U.S. program planned to dispose of 34 metric tons of surplus weapons plutonium in this manner—"enough for thousands of nuclear weapons," according to NNSA.[6] NNSA subsequently proposed making MOX fuel from an additional 7.1 metric tons of plutonium from retired nuclear weapons and another 6 metric tons of weapons-useable plutonium from other sources, for a total of 47.1 metric tons.[7]

The first step in the plutonium disposition process is to disassemble the plutonium "pits," or cores, from nuclear weapons or other sources. Then the plutonium metal must be converted to plutonium oxide to be useable as nuclear reactor fuel. NNSA announced in 2012 that it had produced 442 kilograms of plutonium oxide at Los Alamos National Laboratory in New Mexico and had begun plutonium oxide production at the Savannah River Site.[8]

Under the MOX disposition approach, the plutonium oxide produced at Los Alamos and Savannah River would be blended with uranium, placed in fuel rods, and connected into fuel assemblies for standard commercial power reactors. This fuel fabrication process would take place at the Mixed Oxide Fuel Fabrication Facility. After receiving a construction permit from the Nuclear Regulatory Commission (NRC) in 2005, MFFF began construction in August 2007.[9] MFFF is designed to convert up to 3.5 metric tons of plutonium per year into MOX fuel and would operate for about 20 years.[10]

Getting utilities to commit to loading MOX fuel into their reactors has been a major problem for the plutonium disposition program. Although MOX fuel has been routinely used in Europe, it has enough differences from the conventional uranium fuel currently used by U.S. nuclear plants that it would require separate NRC licensing and special handling. The use of MOX fuel would also probably create public controversy. Two utilities that had previously planned to use the MOX fuel, Dominion and Duke, are no longer involved.

The Tennessee Valley Authority (TVA) subsequently entered into an agreement with DOE "to evaluate the use of mixed oxide fuel in reactors at TVA's Browns Ferry and Sequoyah Nuclear Plants." However, TVA has not yet committed to using MOX fuel, or indicated which of its reactors would be used.[11] The selection of reactors is important, because MOX fuel assemblies must be fabricated to fit particular reactors. Particularly significant are the design differences between fuel assemblies for boiling water reactors, such as Browns Ferry, and pressurized water reactors, such as Sequoyah.

Management and Cost

NNSA's initial February 2002 report to Congress for the MOX plutonium disposition program estimated construction of MFFF would begin in FY2004 and cost about $1 billion to design and build. By July 2005, the design and construction cost estimate had risen to $3.5 billion, according to the DOE

Office of Inspector General (OIG). OIG identified numerous management weaknesses that contributed to the rising costs. However, NNSA contended that much of the cost increase resulted from changes in project scope, such as increases in up-front design work, and delays in Russia over liability issues, which caused parallel delays in the U.S. program.[12]

By the time MFFF construction began in 2007, the construction cost estimate had risen to $4.8 billion. About $100 million of that cost increase was caused by congressional funding holdups, according to NNSA.[13] For example, the Bush Administration requested $653.1 million for fissile materials disposition in FY2006, while Congress appropriated $473.5 million. But for FY2004, in contrast, Congress approved the full request of $656.5 million.

Congress required in 2002 that MFFF produce at least 1 metric ton of MOX fuel by the end of 2008 and complete the then-planned program of 34 metric tons of plutonium conversion by January 1, 2019 (P.L. 107-314, Section 3182). Section 3116 of the National Defense Authorization Act for FY2013 (P.L. 112-239) extended the deadline for starting MOX production to January 1, 2014 (50 U.S.C. 2566).

DOE announced a major reevaluation of the plutonium disposition program in its FY2014 budget submission to Congress. According to the budget request, the MFFF contractor then estimated that the project's total construction cost would rise from $4.8 billion to $7.78 billion, and that construction would not be completed until November 2019.

Moreover, MFFF's operating costs were estimated at $543 million per year, up from $100.5 million estimated in 2002.[14] The Government Accountability Office (GAO) in a February 2014 report said NNSA had provided several major reasons for the most recent cost increases, including DOE's approval of a cost and schedule estimate in 2007 when the overall design was only about 58% complete. Other problems included higher-than-anticipated equipment installation costs, difficulties in finding qualified suppliers, high staff turnover, and further changes in project scope.[15] GAO had found in 2010 that MFFF had experienced consistent delays during the previous two years, partly because of the installation of nearly 4,000 tons of substandard reinforcing bars.[16]

Because of the cost increases, the Obama Administration slowed the project in FY2013 and FY2014 to consider "alternative plutonium disposition strategies."

A Plutonium Disposition Working Group established by the Secretary of Energy began comparing the MOX option with plutonium destruction in fast neutron reactors and non-reactor disposal options, such as blending and spiking with highly radioactive materials.

DOE's FY2015 budget justification says the total lifecycle cost of the MOX disposition option has risen to $30 billion—from $24.2 billion in April 2013.[17] "Due to these increases, the MOX fuel approach is not viable within available resources," according to the DOE justification. "As a result, the MOX project will be placed in cold stand-by while we further study implementation and costs of options to complete the plutonium disposition mission more efficiently."

Total funding for plutonium disposition would drop from $560.3 million in FY2014 to $286.1 million in FY2015, under the Administration's request. No funding would be provided for operations and maintenance related to the MOX Waste Solidification Building, plutonium disposition and infrastructure, and management and integration.

Preparation of plutonium material for disposition would continue. Construction funding for MFFF would be cut from $402.7 million (including reprogrammed funds) to $196.0 million as the facility is placed into "cold standby," or indefinite suspension. Contract personnel at the MOX project declined from 2,271 at the beginning of April 2013 to 1,523 by the end of December, according to the budget justification.[18]

MOX AND NONPROLIFERATION

The extent to which the plutonium disposition programs help or harm nonproliferation goals has been debated since their inception. Some analysts have criticized the MOX option on the principle of opposing any use of plutonium in power generation. From this point of view, nations that do not possess nuclear weapons could use a plutonium-based fuel cycle for power reactors as a cover for developing nuclear weapons. If weapons states such as Russia and the United States used plutonium for power generation, according to this argument, it would be more difficult to persuade non-weapons states not to do so.

Another issue has been whether the Russian disposition program that will make fuel for fast reactors will result in more or less plutonium in the long run. Fast reactors can "breed" or "burn" plutonium as a net output. The reactor acts as a breeder (producing more plutonium than it destroys) if the core includes sufficient amounts of uranium-238, which can absorb a neutron and then decay into the fissile isotope plutonium-239. Without uranium-238, the reactor would consume the original plutonium fuel without producing new plutonium in its place.

The potential for breeding new plutonium was a major concern of the United States in negotiations over the PMDA Protocol and resulted in restrictions on producing new plutonium in these fast reactors using this material. The verification agreement negotiations include discussions of how to technically verify this. In the 2010 Protocol, Russia agreed to restrictions on the reactors that are intended to prevent the creation of more plutonium than is destroyed.[19]

An additional criticism of the program from a nonproliferation perspective is that the high cost required to date for the plutonium disposition program in the United States is funded out of the Defense Nuclear Nonproliferation budget. There is concern that money is being taken away from other programs which some consider to have a more direct impact on preventing the proliferation of weapons-usable nuclear materials. Programs cited include the Second Line of Defense border security program or the Global Threat Reduction Initiative. Non-governmental nonproliferation experts have made recent public statements.[20]

For example, Joe Cirincione of the Ploughshares Fund has said, "funding for the plutonium fuel facility falls within the nuclear nonproliferation budget. That means that every dollar spent on unnecessary programs like MOX is one dollar less for vital nonproliferation programs that keep nuclear material and technology out of the hands of terrorists."[21] In light of this concern, in June 2012 the House passed an amendment to the FY2013 Energy and Water Appropriations bill proposed by Representative Fortenberry to reprogram money from the U.S. MOX program to the Global Threat Reduction Initiative.[22] The provision ultimately was not enacted, however.

Other observers, including within the executive branch, note that the purpose of this program is nonproliferation—rendering weapons-grade plutonium unusable for nuclear weapons on the basis of mutual verification. The American Nuclear Society has come out strongly in favor of the program for nonproliferation reasons.[23]

ISSUES FOR CONGRESS

Recent public debate surrounding the U.S. plutonium disposition program has centered on the budget trade-offs of funding the U.S. plutonium disposition program in light of sequestration and other funding limitations. Local concerns have focused on potential economic losses in the region and the possibility that plutonium brought to Savannah River for use as nuclear fuel will end up being stored there indefinitely. National concerns have been raised that a halt or serious delay to the U.S. program would cause the Russians to reconsider their obligations to dispose of their excess weapons plutonium.

Due to the cost and management challenges of the MOX program, others have raised the possibility of finding an alternative to MOX fuel production as a means to fulfill the U.S.-Russian PMDA.[24] Supporters, however, counter that sunk costs in the current program would make pursuing an alternative path more expensive in the long run.[25]

In testimony before Congress in early 2013, Obama Administration officials reiterated their commitment to the PMDA agreement with Russia.[26] The FY2014 and FY2015 DOE budget justifications state that "NNSA remains committed to the plutonium disposition mission." However, as noted above, the Administration is now considering alternative strategies for achieving that mission.[27]

According to the 2010 Protocol amending the PMDA, the United States would have to obtain written agreement with Russia to implement any alternative to irradiating plutonium in nuclear reactors.[28]

In its report on the National Defense Authorization Act for FY2014 (H.R. 1960, H.Rept. 113- 102), the House Committee on Armed Services expressed concern about MFFF's "continuing escalating costs" but suggested that the Administration's proposed budget cuts "may not actually reduce costs to the taxpayer and will likely delay the disposition of 34 metric tons of weapons grade plutonium."

The Committee directed NNSA to study potential cost savings, consider adding international partners to the program, and study "the potential for achieving greater economic efficiencies by designating additional supplies of surplus plutonium for disposition through the MOX facility." However, the bill as approved by the Committee retained the Administration's proposed construction funding cut.

The Senate Armed Services Committee approved its version of the defense authorization measure (S. 1034) on June 14, 2013. The Committee voted to boost authorized funding for surplus nuclear materials disposition by $80 million over the Administration request.[29] Congress appropriated $526.1 million for Fissile Materials Disposition for FY2014, a cut of $136.7 million from the FY2013 level.

However, the FY2014 appropriation was bolstered by a $59.2 million reprogramming from the International Material Protection and Cooperation account, according to DOE's FY2015 justification.

The State of South Carolina filed a lawsuit on March 18, 2014, to block DOE from placing MFFF in cold standby and to require that construction continue in FY2014. The suit maintains that the Bob Stump National Defense Authorization Act for Fiscal Year 2003 (P.L. 107-314) "was enacted to codify the commitments of the United States and DOE to the State of South Carolina that while plutonium may be placed in South Carolina, such placement was not final disposition for longterm storage of plutonium in the State, but rather a temporary storage to implement the disposition method of MOX processing in the MOX Facility."[30]

Some South Carolina lawmakers and officials have also stressed that South Carolina agreed to accept surplus plutonium from other weapons complex facilities with the understanding that it would be removed under a definite schedule. Plutonium that is made into MOX fuel, they point out, would be shipped off-site to commercial nuclear power plants, while alternatives such as immobilization in glass would provide no clear path for removal, given the current lack of a federal high-level waste central storage facility or underground repository.[31]

Section 3181 of the Bob Stump Defense Authorization (50 U.S.C. 2566) would penalize DOE if it misses various MOX fuel fabrication deadlines. As amended by Section 3116 of the National Defense Authorization Act for 2013 (P.L. 112-239), which delayed the previous deadlines by two years, the act requires that at least one ton of MOX fuel be produced each year, starting in 2014, or plutonium transfers into the state must be halted. If plutonium brought into the state since 2002 is not processed into MOX fuel by 2022, DOE must pay the state up to $100 million per year until it is removed.

DOE said in its FY2015 budget justification that it would not meet the 2014 MOX production deadline in P.L. 112-239. In light of the act's requirements, DOE said that it had "suspended any further transfers of defense plutonium and defense plutonium materials to be processed at the MOX facility in the State of South Carolina."

DOE also said that it "will submit a report to Congress on options for removing an amount of defense plutonium or defense plutonium materials from the State of South Carolina equal to the amount of defense plutonium or defense plutonium materials transferred to the State of South Carolina after April 15, 2002."

Table 2. Defense Nuclear Nonproliferation Appropriations FY2007-FY2015 (in $ thousands)

	FY07	*FY08*	*FY09*	*FY10*	*FY11*	*FY12*	*FY13*	*FY14 Enacted*	*FY14 Repro-gram*[a]	*FY15 Request*
Fissile Materials	470,062	66,235	41,774	701,900	802,198	685,386	663,754	526,057	585,300	311,125
Disposition										
U.S. Surplus	470,062	66,235	40,774	700,900	802,173	684,386	662,832	526,057	585,300	311,125
Materials										
Disposition										
U.S. Pu	57,415	0	0	91,659	200,400	205,632	189,480	157,557	157,557	85,000
Disposition										
U.S. Uranium	86,898	66,235	39,274	34,691	25,985	26,000	23,958	25,000	25,000	25,000
Disposition										
Supporting	14,960	0	1,500	312	0	0	0	0	0	0
Activities										
Construct.	310,789	0	0	574,238	575,788	452,754	449,394	343,500	402,743	201,125
MOX Fuel Fab	262,500	0	0	504,238	501.788	435,172	400,990	343,500	402,743	196,000
Pit Dis.	32,789	0	0	0	17.000	0	0	0	0	0
Waste Solid.	15,500	0	0	70,000	57.000	17,582	48,404	0	0	5,125
Russian Materials	0	0	1,000	1,000	25	1,000	922	0	0	0
Disposition										

Source: CRS-compiled information from DOE annual budget requests to Congress, H.Rept. 112-118, and S.Rept. 112-75.

[a] Reflects a reprogramming of $59,242,760 from FY2013 International Material Protection and Cooperation funding to Fissile Material Disposition in FY 2014.

End Notes

[1] 2000 Plutonium Management and Disposition Agreement as amended by the 2010 Protocol, Article III, http://fissilematerials.org/library/PMDA2010.pdf.

[2] DOE, FY 2015 Congressional Budget Request, Vol. 1, March 2014, pp. 523, 525, 543, http://energy.gov/sites/prod/ files/2014/03/f12/Volume%201%20NNSA.pdf.

[3] This is a sub-program of the NNSA's Defense Nuclear Nonproliferation budget, Fissile Materials Disposition program.

[4] DOE, FY 2015 Congressional Budget Request, Vol. 1, March 2014, p. 536, http://energy.gov/sites/prod/files/2014/03/ f12/Volume%201%20NNSA.pdf.

[5] The reactor's rated thermal power is 2,100 megawatts (MWt), generating 880 megawatts of electricity (MWe), http://www.world-nuclear.org/info/Current-and-Future-Generation/Fast-Neutron-Reactors/.

[6] NNSA, "Plutonium Disposition," http://nnsa.energy.gov/aboutus/ourprograms/nonproliferation/programoffices/ fissilematerialsdisposition/plutoniumdisposition.

[7] NNSA, "Surplus Plutonium Disposition Supplemental Environmental Impact Statement," http://www.nnsa.energy.gov/aboutus/ouroperations/generalcounsel/nepaoverview/nepa/spdsupplementaleis.

[8] NNSA, "NNSA Completes Milestones for Initial Steps in Plutonium Disposition," press release, November 16, 2012, http://nnsa.energy.gov/mediaroom/pressreleases/ oxideprod 111612. A somewhat smaller amount of plutonium metal would be used as feed material for the plutonium oxide, depending on isotopic composition.

[9] NNSA, "NNSA's MOX Fuel Fabrication Facility and U.S. Plutonium Disposition Program," fact sheet, February 14, 2011, http://www.nnsa.energy.gov/mediaroom/factsheets/mox.

[10] Shaw Areva MOX Services, "MOX Fuel Fabrication Facility Construction," http://www.moxproject.com/ construction.

[11] NNSA, "Surplus Plutonium Disposition Supplemental Environmental Impact Statement," op. cit.

[12] DOE Office of Inspector General, Audit Report: Status of the Mixed Oxide Fuel Fabrication Facility, DOE/IG-0713, December 2005, http://energy.gov/sites/prod/files/ igprod /documents/CalendarYear2005/ig-0713.pdf.

[13] Daniel Horner, "Hill-Imposed Delay in MOX Program Costing $115 Million, DOE Says," NuclearFuel, April 23, 2007.

[14] DOE, FY 2007 Congressional Budget, Vol. 1, p. 558, http://energy.gov/sites/ prod/files/FY07Volume1.pdf.

[15] GAO, Plutonium Disposition Program: DOE Needs to Analyze the Root Causes of Cost Increases and Develop Better Cost Estimates, GAO-14-231, February 2014, http://www.gao.gov/assets/670/660927.pdf.

[16] GAO, "Nuclear Nonproliferation: DOE Needs to Address Uncertainties with and Strengthen Independent Safety Oversight of Its Plutonium Disposition Program," GAO-10-378, March 2010, p. 9, http://www.gao.gov/assets/310/ 302279.pdf.

[17] GAO, 2014, op. cit.

[18] DOE, FY 2015 Congressional Budget Request, op. cit

[19] Governments of the United States and the Russian Federation, 2000 Plutonium Management and Disposition Agreement as Amended by the 2010 Protocol, Article III.3, http://fissilematerials.org/library/PMDA2010.pdf.

[20] Reif Kingston, "New Obama Budget Slashes Nonproliferation," The Center for Arms Control and Non-Proliferation, March 6, 2014, http://armscontrolcenter.org/issues/nuclearterrorism/articles/ new_obama_budget_slashes_nonproliferation/.

[21] Joe Cirincione, "The Federal Government's $10 Billion Plutonium Boondoggle," The Atlantic, April 27, 2012, http://www.theatlantic.com/politics/archive/2012/04/the-federal-governments-10-billion-plutonium-boondoggle/ 256470/.

[22] "House Accepts Fortenberry Measure to Strengthen Nuclear Security Efforts," Press Release, Office of Representative Jeff Fortenberry, June 6, 2012, http://fortenberry.house.gov/index.php?option=com_content&view= article&id=3662:house-accepts-fortenberry-measure-to-strengthen-nuclear-security-efforts&catid=41:press-releases& Itemid=2; "House Bill Shifts Money from MOX Plant to Nuclear Security," Global Security Newswire, June 7, 2012.

[23] American Nuclear Society, Position Statement 47, http://www.new.ans. org/pi/ps/docs/ps47.pdf.

[24] Frank von Hippel, Rodney Ewing, Richard Garwin, Alison MacFarlane, "Time to Bury Plutonium," Nature, May 10, 2012. For DOE's analyzed MOX alternatives, see Draft Surplus Plutonium Disposition Supplemental Environmental Impact Statement, summary, DOE/EIS-0283-S2, July 2012, http://nnsa.energy.gov/sites/default/files/nnsa/07-12-inlinefiles/Summary.pdf.

[25] http://us.arevablog.com/2013/02/19/answers-to-questions-about-funding-mox-project-budget/.

[26] House Appropriations Subcommittee on Energy and Water Development Hearing on the Proposed Fiscal 2014 Appropriations for the Energy Department's Nuclear Nonproliferation and Naval Reactor Programs and the National Nuclear Security Administration, February 26, 2013.

[27] DOE, FY 2014 Congressional Budget Request, Vol. 1, April 2013, p. DN-119, http://energy.gov/sites/prod/files/ 2013/04/f0/Volume1.pdf.

[28] 2000 Plutonium Management and Disposition Agreement as amended by the 2010 Protocol, Article III, http://fissilematerials.org/library/PMDA2010.pdf.

[29] Senate Committee on Armed Services, "Senate Committee on Armed Services Completes Markup of the National Defense Authorization Act for Fiscal Year 2014," June 14, 2013, http://www.armed-services.senate.gov/press/releases/ upload/SASC_NDAA_061413.pdf.

[30] State of South Carolina v. United States Department of Energy, United States District Court, District of South Carolina, Aiken Division, March 18, 2014, http://2hsvz0l74ah31vgcm16peuy12tz.wpengine.netdna-cdn.com/wpcontent/uploads/2014/03/MOX-complain-3.18.2014-00269474xD2C78.pdf.

[31] Rob Pavey, "Sen. Lindsey Graham Says Reducing MOX Costs Could Save the Project," Augusta Chronicle, April 25, 2013, http://chronicle.augusta.com/news/government/2013-04-25/sen-lindsey-graham-says-reducing-mox-costscould-save-project.

In: Plutonium Disposition
Editor: Glenn Reiss
ISBN: 978-1-63321-066-0

Chapter 2

Plutonium Disposition Program: DOE Needs to Analyze the Root Causes of Cost Increases and Develop Better Cost Estimates*

United States Government Accountability Office

Why GAO Did This Study

NNSA, a separately organized agency within DOE, manages the Plutonium Disposition program to dispose of surplus weapons-grade plutonium by burning it as MOX fuel—a mixture of plutonium and uranium oxides—in specially modified commercial nuclear reactors. In 2012, DOE forecasted cost increases of close to $3 billion over the previous estimates for the program's two construction projects, the MOX facility and the WSB for disposing of waste from the MOX facility.

GAO was asked to review these cost increases and the life-cycle cost estimate. This report examines: (1) drivers NNSA identified for the cost increases; (2) the extent to which NNSA analyzed underlying causes of the cost increases; (3) steps NNSA took to hold construction contractors accountable for their role, if any, in the cost increases; and (4) the extent to

* This is an edited, reformatted and augmented version of United States Government Accountability Office Publication, No. GAO-14-231, dated February 2014.

which NNSA's most recent estimates met cost- and schedule-estimating best practices. GAO reviewed NNSA's draft life-cycle cost estimate and contractor estimates of the MOX project's cost and WSB schedule, compared the estimates with cost- and schedule-estimating best practices, and interviewed DOE and NNSA officials.

WHAT GAO RECOMMENDS

GAO is recommending, among other things, that DOE conduct a root cause analysis of the Plutonium Disposition program's cost increases and ensure that future estimates of the program's life-cycle cost and cost and schedule for the program's construction projects meet all best practices for reliable estimates. DOE generally agreed with GAO's recommendations.

WHAT GAO FOUND

The Department of Energy's (DOE) National Nuclear Security Administration (NNSA) identified various drivers for the close to $3 billion increase in the estimated cost of the Plutonium Disposition program's two construction projects—the Mixed Oxide (MOX) Fuel Fabrication Facility and the Waste Solidification Building (WSB). These drivers included DOE's approval of the MOX facility's cost and schedule estimates before design was complete and schedule delays in construction of the WSB. According to NNSA, the cost of critical system components for the MOX facility averaged 60 percent higher than estimated as a result of approval of estimates before design was complete.

NNSA has not analyzed the underlying, or root, causes of the Plutonium Disposition program construction cost increases to help identify lessons learned and help address the agency's difficulty in completing projects within cost and schedule, which has led to NNSA's management of major projects remaining on GAO's list of areas at high risk of fraud, waste, abuse, and mismanagement. DOE's project management order requires that lessons learned be captured throughout a project to, among other things, benefit future endeavors. NNSA officials said that, because the order does not require a root cause analysis of cost increases, NNSA decides on a case-by-case basis whether to conduct one. Unlike a root cause analysis, the cost drivers NNSA identified provided few details about why the drivers existed, such as DOE's reasons for approving the MOX facility's cost and schedule estimates before the design was complete. Without a root cause analysis, it is uncertain whether NNSA will be able to accurately identify underlying causes of the increases to identify and implement corrective measures and identify lessons learned to apply to other projects.

After determining that the performance of the contractors for the MOX facility and WSB contributed to cost increases, NNSA took steps to hold the contractors accountable by withholding fees specified under the contracts. In particular, as of November 2013, NNSA withheld $45.1 million or close to one-third of the MOX contractor's fees, including fees tied to meeting the MOX project's cost and schedule estimates. In addition, NNSA withheld $7.7 million or about 40 percent of the WSB contractor's fees tied to various performance measures for the WSB, such as completing construction milestones.

NNSA's most recent estimates for the Plutonium Disposition program did not fully reflect all the characteristics of reliable cost estimates (e.g., credible) and schedule estimates (e.g., well-constructed) as established by best practices for cost- and schedule-estimating, placing the program at risk of further cost increases. For example: (1) NNSA's draft April 2013 life-cycle cost estimate of $24.2 billion for the overall program was not credible because NNSA did not conduct an independent cost estimate to provide an unbiased test of whether the estimate was reasonable. (2) Because the MOX contractor's September 2012 proposal for increasing the cost of the MOX facility did not include a formal analysis to examine the effects of changing assumptions, it was minimally credible. (3) The WSB contractor's February 2013 monthly update to its schedule estimate was minimally well-constructed in that it contained activities that were not properly tied with the start or end date of other activities, which could potentially obscure the critical path determining the project's completion date.

ABBREVIATIONS

DOE	Department of Energy
EVM	earned value management
FAR	Federal Acquisition Regulation
MIFT	MOX Irradiation, Feedstock, and Transportation
MOX	mixed oxide

NNSA	National Nuclear Security Administration
NRC	Nuclear Regulatory Commission
PDCF	Pit Disassembly and Conversion Facility
PDIP	Plutonium Disposition and Infrastructure Program
WSB	Waste Solidification Building

February 13, 2014
The Honorable Mike Simpson
Chairman

The Honorable Marcy Kaptur
Ranking Member
Subcommittee on Energy and Water Development, and Related Agencies
Committee on Appropriations
House of Representatives

Plutonium is a man-made, radioactive element that poses a danger of nuclear weapons proliferation and a risk to human health and the environment. Under an agreement signed in 2000, the United States and Russia have each pledged to dispose of at least 34 metric tons of surplus weapons-grade plutonium no longer needed for defense purposes by burning it as mixed oxide (MOX) fuel in specially modified commercial nuclear reactors.[1] Once used and removed from a reactor, the plutonium can no longer be readily used to make a nuclear weapon. The National Nuclear Security Administration (NNSA), a separately organized agency within the Department of Energy (DOE), manages the Plutonium Disposition program. As part of this program, NNSA entered into separate contracts for design and construction services for two facilities at DOE's Savannah River Site in South Carolina: (1) the MOX Fuel Fabrication Facility for producing MOX fuel for nuclear reactors and (2) the Waste Solidification Building (WSB) for disposing of liquid waste from the MOX facility.[2]

Under DOE's project management order[3] and related policies and guidance, NNSA project directors are responsible for managing the MOX facility and WSB construction projects and overseeing the contractors that design and construct the facilities. Among other things, the project management order establishes a process for DOE and NNSA to review and approve a project's construction cost and schedule estimates, including changes to the estimates; report monthly on a project's cost and schedule performance; and conduct project reviews at least once per year.

In February of 2013, NNSA's contract administration and project management were again included on GAO's list of areas at high risk of fraud, waste, abuse, and mismanagement.[4] In particular, NNSA has experienced long-standing difficulties in completing major projects within cost and on schedule. As we testified in March 2013, DOE forecasted an increase in its estimated cost and schedule to complete the MOX facility and WSB.[5] Specifically, in 2012, DOE forecasted a close to $3 billion increase in the estimated cost to complete the two facilities:

- *MOX facility*. In April 2007, DOE approved a cost estimate for the MOX facility of $4.8 billion and start of operations in September 2016.[6] Construction began in August 2007. In 2012, at NNSA's direction to update the estimate, the MOX contractor submitted a proposal to increase the cost of the facility to $7.7 billion with the start of operations delayed to November 2019. DOE began evaluating the proposal and, pending the outcome of its evaluation, directed the contractor to use its proposed cost and schedule estimate as a provisional baseline for purposes of monthly reporting.
- *WSB*. In December 2008, NNSA approved start of construction of the WSB and a cost estimate of $344.5 million and start of operations in September 2013.[7] In February 2012, NNSA directed the WSB contractor to develop a plan to address potential cost and schedule overruns. In December 2012, NNSA approved an increase in the estimated cost to $414.1 million and a delay in the start of operations to August 2015.

NNSA has not finalized a life-cycle cost estimate for the Plutonium Disposition program—that is, an estimate of all costs to complete the mission to dispose of surplus weapons-grade plutonium—but, in April 2013, it completed a draft estimate of $24.2 billion. In addition to construction costs, the estimate included operation and maintenance of the MOX facility and WSB, as well as other components of the program, such as production of plutonium feedstock for the MOX facility. The life-cycle cost estimate also included $5.2 billion in actual costs through fiscal year 2012: $3.4 billion for the MOX facility, $265 million for the WSB, and $1.5 billion for other components of the program.

In 2013, in light of the cost increases for the Plutonium Disposition program's construction projects, the fiscal year 2014 budget request for NNSA stated that converting plutonium to MOX fuel may be unaffordable. The budget request announced that, as a result, NNSA would slow down activities associated with the current plutonium disposition strategy during an assessment of alternative plutonium disposition strategies. As of January 2014, this assessment had not yet been completed.

To provide assistance to federal agencies in preparing cost and schedule estimates, we have compiled best practices used throughout government and industry. In March 2009 and May 2012, we issued guides identifying the characteristics of high-quality, reliable cost and schedule estimates, respectively.[8] Specifically, the four characteristics of a high-quality cost estimate are comprehensive, well-documented, accurate, and credible, and the four characteristics of a high-quality schedule estimate are comprehensive, well-constructed, credible, and controlled.

You asked us to review issues related to the construction cost increases for the MOX facility and WSB and the life-cycle cost of the overall Plutonium Disposition program. This report examines: (1) drivers of the cost increases that NNSA identified for the Plutonium Disposition program's construction projects, (2) the extent to which NNSA analyzed underlying causes of the cost increases, (3) steps NNSA took under the contracts for the construction projects to hold the contractors accountable for their role, if any, in the cost increases, and (4) the extent to which NNSA's most recent estimates for the program's life-cycle cost and the cost and schedule for completing the program's construction projects met best practices for reliable cost and schedule estimates.

To assess drivers of the cost increases, we reviewed the MOX contractor's September 2012 proposal for increasing the project's cost, the change to the WSB project's cost approved in December 2012, and other documents. In addition, we visited the Savannah River Site to observe construction progress for both projects and interviewed NNSA and contractor officials. Because the MOX facility represents most of the program's construction cost increase, we also analyzed the MOX contractor's system for tracking and reporting on cost and schedule performance. To determine the extent to which NNSA analyzed underlying causes of the cost increases, we reviewed DOE and NNSA documentation of the cost increases and interviewed NNSA officials. To determine steps taken to hold contractors accountable for their role in the cost increases, we reviewed the contracts for both projects, fees specified under the contracts, and NNSA's fee evaluations and other documentation supporting its fee determinations. In addition, we obtained NNSA data on fees it paid to and withheld from the MOX and WSB contractors. We assessed the reliability of the data by interviewing NNSA contracting officers responsible for administering the contracts and other means, and we determined that the data were sufficiently reliable for reporting on the fees NNSA paid and withheld.

To assess the extent to which NNSA's most recent estimates of the program's life-cycle cost and the cost and schedule for completing the program's construction projects met best practices, we tailored our methodology to the differing stages of NNSA's development and approval of each estimate. Specifically, because NNSA had not finalized a life-cycle cost estimate for the program or a revised cost and schedule estimate for the MOX facility, we assessed the most recent available estimates—spreadsheets dated April 2013 representing NNSA's draft life-cycle cost estimate and the MOX contractor's September 2012 proposal for increasing the project's cost. We assessed the WSB schedule estimate because, as described in the *GAO Schedule Assessment Guide*, a reliable schedule can contribute to an understanding of the cost impact if a project does not finish on time. Specifically, we compared the contractor's February 2013 monthly update to its schedule estimate, which was the most recent available update when we conducted our analysis. Additional details on our objectives, scope, and methodology can be found in appendix I.

We conducted this performance audit from November 2012 to February 2014 in accordance with generally accepted government auditing standards. Those standards require that we plan and perform the audit to obtain sufficient, appropriate evidence to provide a reasonable basis for our findings and conclusions based on our audit objectives. We believe that the evidence obtained provides a reasonable basis for our findings and conclusions based on our audit objectives.

Background

The end of the cold war left the United States with a surplus of weapons-grade plutonium. Much of this material is found in a key nuclear weapon component known as a pit. In 1997, DOE announced a plan to dispose of surplus, weapons-grade plutonium through an approach that included fabrication of plutonium into MOX fuel for use in domestic commercial nuclear reactors. In 2000, the United States and Russia entered into a Plutonium Management and Disposition Agreement, in which each country pledged to dispose of at least 34 metric tons of surplus, weapons-grade plutonium. Through a protocol to the agreement signed in 2010, the United States and Russia reaffirmed their commitment to dispose of surplus, weapons-grade plutonium as MOX fuel in nuclear reactors, and the agreement entered into force in 2011.

The MOX facility is designed to remove impurities from plutonium feedstock obtained from nuclear weapon pits, form the plutonium into MOX fuel pellets, and fabricate pellets into fuel assemblies for use in a reactor. The MOX facility is a reinforced concrete structure measuring about 600,000 square feet (including support buildings) and, when complete, will include about 300 separate process systems using approximately 23,000 instruments; 85 miles of process piping; 500,000 linear feet of conduit; 3,600,000 linear feet of power and control cable; and 1,000 tons of heating, ventilation, and air conditioning duct work. The WSB will be a 33,000 square foot reinforced concrete structure and will include tanks, evaporators, and solidification equipment to process radioactive liquid waste streams from the MOX facility into solid waste forms suitable for disposal at DOE sites in New Mexico and Nevada. Figure 1 shows aerial views of construction progress for the MOX facility and WSB as of June 2013 and July 2013, respectively.

MOX Fuel Fabrication Facility.

Figure 1. (Continued).

Waste Solidification Building.

Source: National Nuclear Security Administration.

Figure 1. Aerial Views of the Plutonium Disposition Program's Construction Projects.

In addition to the MOX facility and WSB, NNSA's plans for the U.S. Plutonium Disposition program include the following two additional components:

- *MOX Irradiation, Feedstock, and Transportation (MIFT).* Among other activities, this component includes: (1) production of plutonium feedstock for the MOX facility, (2) qualification of MOX fuel for use in commercial nuclear reactors, and (3) procurement and maintenance of shipping containers for plutonium feedstock and MOX fuel.
- *Plutonium Disposition and Infrastructure Program (PDIP).* This component includes overall management and integration of the MOX facility and WSB projects and integration of the projects with activities falling under MIFT; preparation of environmental impact statements and records of decision for the program in accordance with the National Environmental Policy Act; support for infrastructure at the Savannah River Site, such as site roads; and other activities.

NNSA's plans for producing plutonium feedstock previously included design and construction of a stand-alone Pit Disassembly and Conversion

Facility (PDCF) at the Savannah River Site. As we reported in March 2010, NNSA never established a definitive cost and schedule estimate for the PDCF,[9] but NNSA estimated in January 2011 that the cost of the facility could range from $4.5 billion to $4.8 billion. NNSA canceled the PDCF in January 2012 and, instead, proposed in a July 2012 draft environmental impact statement to meet the feedstock requirements for the MOX facility through existing facilities at DOE's Los Alamos National Laboratory and the Savannah River Site.[10] According to NNSA's draft life-cycle cost estimate for the Plutonium Disposition program, NNSA spent $730.1 million on the PDCF prior to its cancellation.

In July 2012, NNSA also announced its preferred alternative for disposition of 13.1 metric tons of surplus plutonium not already included in the 34 metric tons planned for disposal as MOX fuel. The additional plutonium included pits declared excess to national defense needs, as well as surplus non-pit plutonium. According to NNSA officials, the preferred alternative would increase the amount of plutonium disposed as MOX fuel to about 42 metric tons. As of December 2013, DOE had not issued a final supplemental environmental impact statement or record of decision on the facilities to be used to meet plutonium feedstock requirements for the MOX facility or on the disposition pathway for the 13.1 metric tons of surplus plutonium.

NNSA's Office of Defense Nuclear Nonproliferation provides policy direction for the Plutonium Disposition program, develops and manages annual budgets and the life-cycle cost estimate for the overall program, and manages the MIFT and PDIP components of the program.

NNSA's Office of Acquisition and Project Management is responsible for managing construction of the MOX facility and WSB projects within approved cost and schedule estimates. To do so, the office manages teams of federal project directors and federal staff that provide direction and oversight of the contractors for both projects, report monthly on the projects' cost and schedule performance, and evaluate contractors' performance in areas such as management of subcontractors. The office also conducts reviews of the construction projects to evaluate technical, cost, scope, and other aspects of the projects so that any necessary course corrections can be made. DOE's project management order requires that such reviews be conducted at least once per year.

NNSA entered into cost-reimbursable contracts for construction of the MOX facility and WSB. A cost-reimbursable contract provides for payment of a contractor's allowable incurred costs to the extent prescribed in the contract.[11] Agencies may use cost-reimbursable contracts when uncertainties in the scope of work or cost of services prevent the use of contract types in which prices are fixed, known as fixed-price contracts. The MOX and WSB contracts included fees with payment tied to meeting or exceeding preestablished requirements or withholding of fees for any requirements not met, thereby reducing contractors' profits. Under the MOX contract, NNSA provided four types of fees that the contractor could earn: (1) incentive fees—a type of fee specifically tied to meeting a project's cost and schedule estimate; (2) milestone fees tied to on-time completion of construction milestones; (3) award fees, which are generally intended to motivate performance in areas other than cost and schedule, such as safety; and (4) fixed fees, a set amount a contractor receives for contract performance. In contrast, NNSA included only one type of fee for the WSB—a performance incentive fee under the contract for management and operation of the Savannah River Site, which included construction of the WSB. In order to provide the contractor performance incentives specifically related to construction of the WSB, NNSA established various performance measures, such as meeting the project's cost and schedule and completing construction milestones, and allocated portions of the fee to each performance measure.

The contractors for the MOX facility and WSB work with subcontractors to construct the facilities. For example, the WSB contractor entered into a subcontract that included all construction activities for the WSB with the exception of early site work, such as installation of underground utilities.

Once the construction subcontractor completes its work, the WSB contractor is responsible for start-up testing and operation of the facility.

Under DOE's project management order, the Deputy Secretary of Energy is the senior DOE official accountable for all of the department's project acquisitions. In addition, the Deputy Secretary approves cost and schedule estimates for all major construction projects—defined as those with values of at least $750 million, which includes the MOX facility—and approves any cost increase over $100 million for a major or nonmajor project. The DOE Office of Acquisition and Project Management conducts external independent reviews to validate estimates prior to approval by the Deputy Secretary. Once estimates have been approved, this office monitors projects' cost and schedule performance and reports to the Deputy Secretary on a monthly basis. Figure 2 depicts the roles of NNSA, DOE, and contractors in managing the Plutonium Disposition program.

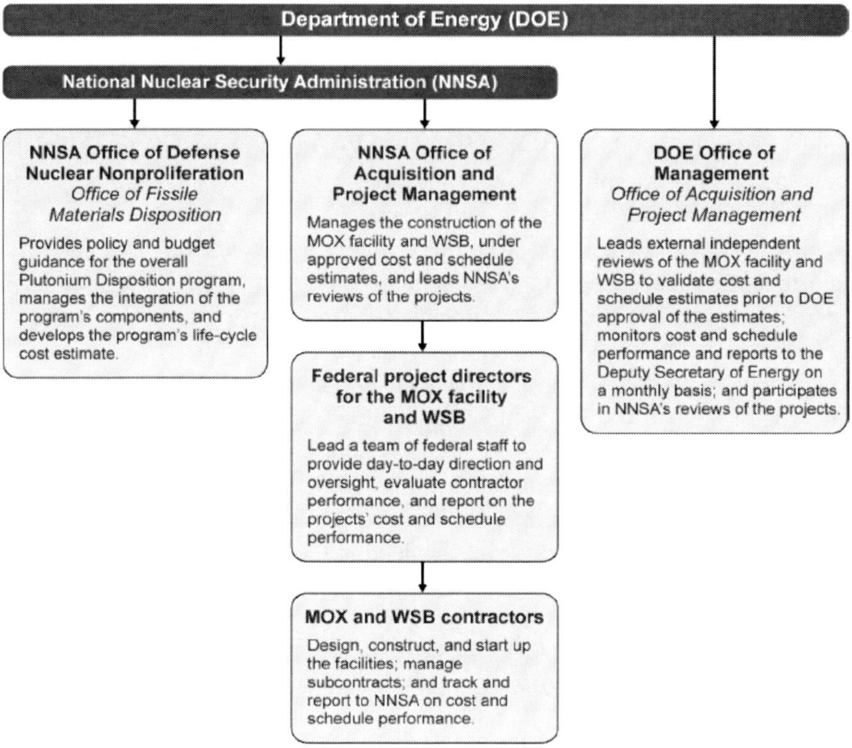

Source: GAO.

Figure 2. DOE and NNSA Organizational Structure for Management of the Plutonium Disposition Program.

The *GAO Cost Estimating and Assessment Guide*[12] and the *GAO Schedule Assessment Guide*[13] compiled best practices corresponding to the four characteristics of high-quality, reliable cost and schedule estimates, respectively:

- The characteristics of a high-quality, reliable cost estimate are comprehensive, well-documented, accurate, and credible. For example, (1) a comprehensive estimate has enough detail to ensure that cost elements are neither omitted nor double counted, (2) a well-documented estimate allows for data it contains to be traced to source documents, (3) an accurate estimate is based on an assessment of most likely costs and has been adjusted properly for inflation, and (4) a credible estimate discusses any limitations because of uncertainty or bias surrounding data or assumptions. Our cost estimating guide also lays out 12 key steps that should result in high-quality cost estimates. For example, one of the steps is to conduct an independent cost estimate—that is, one generated by an entity that has no stake in approval of the project but uses the same detailed technical information as the project estimate. Having an independent entity perform such a cost estimate and comparing it with a project team's estimate provides an unbiased test of whether a project team's estimate is reasonable.

- The four characteristics of a high-quality, reliable schedule are comprehensive, well-constructed, credible, and controlled. For example, (1) a comprehensive schedule includes all government and contractor activities necessary to accomplish a project's objectives, (2) a well-constructed schedule sequences all activities using the most straightforward logic possible, (3) a credible schedule uses data about risks and opportunities to predict a level of confidence in meeting the completion date, and (4) a controlled schedule is updated periodically to realistically forecast dates for activities.

NNSA Identified Various Cost Drivers for the Plutonium Disposition Program's Construction Projects

NNSA identified various drivers of the cost increases for the MOX facility and WSB. NNSA's budget request for fiscal year 2014 summarized the cost drivers that NNSA considered to be most significant. In addition, NNSA identified some of these drivers in earlier documents, including in reports of project reviews conducted in 2011 and 2012, in monthly status reports for the projects, and, for the WSB, in the document requesting approval for a cost increase. NNSA and contractor officials provided additional details on these drivers during interviews with us.

Key drivers NNSA identified for the cost increase for the MOX facility included the following:

- *DOE's approval of the cost and schedule before design was complete.* The head of NNSA's Office of Acquisition and Project Management told us that, judging from the MOX contractor's design costs during construction of the MOX facility, the overall design was about 58 percent complete when DOE approved the project's cost and schedule estimate in April 2007. In contrast, according to DOE's project management order, to support the development of a cost estimate, the design of complex nuclear processing facilities needs to be closer to 100 percent complete than the design of basic facilities, such as administrative buildings and general purpose laboratories. NNSA's budget request for fiscal year 2014 stated that the cost of critical system components for the MOX facility averaged 60 percent higher than estimated as a result of approval of these estimates before design was complete. According to NNSA and MOX contractor officials, after the contractor completed designs for critical system components, such as the gloveboxes used in the facility for handling plutonium and related infrastructure, equipment suppliers submitted higher bids than the contractor anticipated. For example, according to the contractor's Vice President of Operations, a vendor submitted a bid in 2008 that was four times the amount the same vendor had estimated in 2005.

- *Higher-than-anticipated costs to install equipment.* For example, the MOX contractor estimated in its September 2012 proposal to increase the cost of the facility that the labor hours to install each foot of the approximately 85 miles of piping in the facility increased by as much as 26 percent and that, as facility designs became more definitive, the total amount of pipe increased by close to 33 percent over the previous estimate. In addition, according to NNSA, the number of safety systems needed to meet Nuclear Regulatory Commission (NRC) requirements was greater than anticipated, further adding to equipment installation costs.[14] According to NNSA officials, NNSA and the contractor did not have a good understanding of the cost of designing the facility to meet NRC requirements related to demonstrating the ability to withstand an earthquake. The officials explained that the facility's design is based on a similar facility in France but that NRC regulatory requirements differ from those in France.
- *The contractor's difficulty identifying suppliers and subcontractors able to fabricate and install equipment meeting nuclear quality assurance criteria.*[15] According to NNSA's review of the MOX project in 2011, the project was experiencing the same issues identifying qualified suppliers and subcontractors as other nuclear

projects across DOE. These issues included a higher than expected effort associated with attracting qualified vendors and, after vendors were selected, responding to questions or correcting noncompliance with requirements. For example, according to NNSA and the MOX contractor, the contractor needed to station quality assurance personnel at supplier and subcontractor locations to oversee activities.

- *Greater-than-expected turnover of engineering and technical staff.* In particular, the project lost staff to other nuclear industry projects, including projects in neighboring states, resulting in a nearly complete turnover of construction management personnel over a period of several years and the need to provide training to replacement personnel. NNSA identified this driver in its budget request for fiscal year 2012. Specifically, the budget request stated that over 15 percent of the project's engineering and technical personnel had left for other nuclear industry jobs in the previous year with pay increases of at least 25 percent. The budget request further stated that finding experienced replacements had become difficult and expensive. According to the budget requests for fiscal years 2013 and 2014, the loss of experienced engineering and technical staff to other nuclear industry projects has continued.
- *Change in scope of the project to add capability to the MOX facility to produce plutonium feedstock.* As part of its decision to cancel plans for a stand-alone PDCF and to instead meet feedstock requirements through existing facilities, NNSA directed the MOX contractor to include feedstock capability in its September 2012 proposal to increase the cost of the facility. The contractor's proposal included an estimate of $262.3 million to add feedstock capability.

In identifying these drivers of the cost increase for the MOX facility, NNSA did not identify the dollar amount associated with each cost driver. An NNSA official said that the MOX contractor's system for tracking and reporting on cost and schedule performance could potentially be used to determine dollar amounts that each driver added to the overall cost increase—which is one possible use of such a system—but that doing so would be time-consuming and difficult. As a result, NNSA officials could not substantiate the relative importance of the cost drivers. For example, NNSA officials said they had not conducted a formal analysis to back up an estimate, which they had made when we first discussed the cost drivers with them, that lack of design maturity of critical system components accounted for more than half of the increase. In reviewing the MOX contractor's system, we found that, as NNSA officials stated, using the system to determine the dollar amounts each driver added to the cost increase would be difficult—for example, because the system's identification of cost increases at a summary level, such as site construction support, did not correspond to the cost drivers identified by NNSA.

Key cost drivers NNSA identified for the WSB included the following:

- *Higher-than-anticipated bids for the construction subcontract.*[16] According to the NNSA federal project director for the WSB, the WSB contractor received two bids in 2009 from prospective construction subcontractors that both came in at about $26 million higher than the contractor's estimate. NNSA officials did not explain the reason for the difference, stating that the bidders were not required to provide details of their estimates. The federal project director said that NNSA supported the WSB contractor awarding the construction subcontract, despite the higher cost, in order to maintain the schedule for completing the WSB in time to support the start-up of the MOX facility. According to NNSA officials, the project applied cost savings from earlier work to cover part of the increased cost of the construction subcontract and had sufficient contingency—the portion of a project's budget that is available to account for uncertainties in the project's scope—to absorb the remainder of the increase. Consequently, however, contingency to absorb further cost increases as construction progressed was reduced.
- *Design errors, omissions, and inconsistencies.* According to the NNSA federal project director, the WSB contractor and subcontractor made hundreds of design changes, which led to an additional cost increase in the construction subcontract. According to NNSA's log of design changes, as of August 2013, design changes increased the cost of the construction subcontract by about $15 million, from $91.5 million to $106.5 million. The federal project director said that, unlike the design of the MOX facility, the design of the WSB was about 90 percent complete at the start of construction. A September 2008 report of NNSA's independent review of the WSB prior to approval of the cost and schedule estimate found that the design was essentially complete. Nevertheless, according to the federal project director, design changes were needed because of constructability issues, such

as equipment that met specifications in design documents not being available by the time the project reached construction.

- *Schedule delays resulting from the construction subcontractor not meeting required targets.* According to the NNSA federal project director's feedback on the WSB contractor's performance in September 2009, NNSA had concerns related to the project schedule and the ability to meet the completion date in part because of a delayed start in the construction subcontract. By the time NNSA approved the cost increase for the WSB in December 2012, schedule delays in the construction subcontract had grown to 15 months. The approved cost increase included about $30 million in the contractor's delay-related costs because NNSA's contract for the WSB is cost-reimbursable.

The actual cost attributable to the WSB may be even higher depending on the outcome of a lawsuit filed by the subcontractor against the WSB contractor related to design changes and schedule delays that increased the subcontractor's costs in excess of the amount specified in its fixed-price subcontract. The approved cost increase for the WSB included contingency to account for the possibility of higher costs incurred by the construction subcontractor.

NNSA Has Not Analyzed Root Causes of the Cost Increases for the Plutonium Disposition Program's Construction Projects

NNSA has not analyzed the underlying, or root, causes of the close to $3 billion in construction cost increases for the MOX facility and WSB.[17] DOE's project management order requires that lessons learned be captured throughout a project to allow for the exchange of information within DOE in the context of project management and to benefit future endeavors. However, the project management order does not include a requirement for a root cause analysis of projects experiencing significant cost increases or schedule delays. NNSA officials said that they decide on a case-by-case basis whether to conduct a root cause analysis.[18] In contrast, under the Weapon Systems Acquisition Reform Act of 2009, the Department of Defense must perform a root cause analysis of a cost increase that exceeds a certain threshold.

Documentation NNSA provided to us on the cost drivers for the MOX facility and WSB do not provide clear details about the causes of the cost increases. Such details can be found in a root cause analysis, which would help address questions about why the drivers identified by NNSA occurred and help inform lessons learned. Key questions about the cause of the key drivers include the following:

- DOE's reasons for approving a cost and schedule estimate for the MOX facility before the design was complete, even though a July 2006 review of the project found that the cost estimate's basis on portions of the design that were less than 50 percent complete posed a risk to the project. Similarly, a root cause analysis would address why one of the drivers of the cost increase for the WSB identified by NNSA was design errors, omissions, and inconsistencies, given that a review prior to approval of the project's cost and schedule estimate found that most of the design was ready for construction.
- The extent to which NNSA and its contractors shared responsibility for cost drivers, such as the greater-than-anticipated number of safety systems needed in the MOX facility to meet NRC requirements. According to NNSA officials, the department hired the MOX contractor because it considered the contractor to be well-qualified to engineer and estimate all of the safety systems for the facility, taking into account NRC requirements. However, the record for DOE's approval of the cost and schedule estimate for the facility shows that DOE was aware of complexities in adapting MOX technology to comply with NRC requirements. Specifically, the minutes from DOE's July 2006 meeting to request approval of the estimate stated that these complexities had already contributed to a $1.1 billion increase in the estimated cost.
- The sufficiency of measures DOE took to ensure that the cost estimate for the MOX facility it approved in 2007 reflected an awareness of market conditions, such as the availability of suppliers and subcontractors with the ability and experience to meet nuclear quality assurance criteria. As required under the MOX contract, in October 2006—before DOE approved the cost and schedule estimate for the facility—the contractor submitted a construction market analysis report, which stated that the contractor had experienced trouble obtaining qualified suppliers and that the subcontractor pool using nuclear quality standards had been decreasing due to inactivity in the

nuclear industry. However, the report provided limited detail and did not include recommendations to address availability of qualified suppliers.

- The thoroughness of DOE's review, required under DOE's project management order, to ensure that the WSB contractor's system for tracking and reporting on cost and schedule performance provided accurate information. DOE recertified the contractor's system in December 2011 after identifying and closing out several corrective actions and continuous improvement opportunities. However, DOE found additional problems with the system after January 2012, when the WSB contractor informed NNSA that schedule delays for the project were greater than the contractor previously revealed. Based in part on the contractor's revelations, DOE reexamined the contractor's system and suspended its certification in November 2012.
- The corrective actions NNSA and its contractors took after periodic project reviews identified problems, including problems cited by NNSA as drivers of cost increases for the MOX facility and WSB. For example, multiple reviews of the MOX facility found that costs to install equipment were underestimated. A July 2006 review found that installation for electrical; piping; and heating, ventilation, and air-conditioning equipment were underestimated by close to $160 million and nearly 3 million labor hours. NNSA's project reviews of the facility in 2011 and 2012 continued to raise concerns about unrealistic installation rates.
- The responsiveness of NNSA project managers to emerging cost and schedule issues. Without a review of the timing of NNSA initiating the process of increasing the projects' cost and schedule estimates, it is not clear whether NNSA acted in a timely manner or whether project cost and schedule indicators warranted earlier action. For example, an NNSA review of the MOX facility in the spring of 2011 found that the most significant risk to delivering the project within cost centered on the ability of the project team to identify about $364 million in savings to offset expected cost growth, but NNSA did not initiate the process of increasing the project's cost and schedule estimates until January 2012.

Table 1. Types of Fees for Construction of the MOX Facility and Amounts Paid, Withheld, and Remaining to Be Earned as of November 2013

Dollars in millions

Fee type	Purpose	Total available to be earned under the MOX contract	Amount paid	Amount withheld	Percentage withheld	Amount remainingunder the MOXcontract[a]
Incentive[b]	Encourage project completion within cost and schedule	$76.9	$29.1	$36.5	56%	$11.3
Award[c]	Ensure performance in areas other than cost and schedule	53.1	24.0	8.6	26	20.5
Milestone	Motivate completion of critical tasks	61.0	30.8	0	0	30.2
Fixed	Reward the contractor for work during contract negotiations	15.7	15.7	0	0	0
Total		**$206.6**	**$99.6**	**$45.1**	**31%**	**$62.0**

Source: GAO analysis of NNSA records of fee payments to the MOX contractor.

Note: Dollar figures do not add up to $206.6 million due to rounding.

[a]The MOX contractor's September 2012 proposal for increasing the cost of the MOX facility included increases to incentive, milestone, and award fees specified under the contract.

[b]The amount of remaining incentive fees covers fiscal years 2014 and 2015. Of the $29.1 million in incentive fees paid to the contractor, $21.6 million remains provisional, meaning that NNSA can require that the fees be paid back as a result of the project not being completed within cost. In addition, under the terms of the MOX contract, the contractor could earn the $36.5 million in withheld incentive fees if it completes the overall project within cost and schedule.

[c]NNSA made award fee payments for fiscal years 2008 through 2012, increasing the amount withheld from 11 percent in fiscal year 2008 to 50 percent in fiscal year 2012. The amount remaining covers fiscal years 2013 through 2016.

Without a root cause analysis, it is uncertain whether NNSA will be able to accurately identify underlying causes of the cost increases for the MOX facility and WSB in order to identify and implement corrective measures and identify lessons learned to share with and apply to other DOE construction projects.

NNSA HAS TAKEN STEPS TO HOLD CONTRACTORS ACCOUNTABLE FOR COST INCREASES BY WITHHOLDING FEES

After determining that the performance of the contractors for the MOX facility and WSB contributed to the projects' construction cost increases, NNSA took steps to hold the contractors accountable for their performance by withholding fees specified under the contracts. Specifically, NNSA withheld portions of two of the four types of the MOX contractor's fees and 41 percent of the WSB contractor's fees.

NNSA Withheld Portions of Two of the Four Types of the MOX Contractor's Fees

NNSA withheld portions of two of the four types of fees that the MOX contractor could earn under the contract for construction of the facility—incentive fees and award fees. In total, NNSA withheld $45.1 million or close to one-third of all fees the contractor could earn as of November 2013. Under the terms of the MOX contract, the contractor could still earn incentive fees that have been withheld, but only if it completes the overall project within cost and schedule. Table 1 summarizes fees paid to and withheld from the contractor as of November 2013. Details of fees NNSA withheld and paid under the MOX contract include the following:

- *Incentive fees*. NNSA did not pay $36.5 million or over half of the $65.6 million in incentive fees that the MOX contractor could earn from fiscal year 2008, when construction began, through fiscal year 2013. Of the $29.1 million in incentive fees paid to the contractor, $21.6 million remains provisional, meaning that NNSA can require that the fees be paid back as a result of the project not being completed within cost.[19] The amount not paid represented the contractor's entire incentive fees for fiscal years 2011 through 2013. Specifically, under the terms of the MOX contract, NNSA can withhold quarterly payments of incentive fees if an increase in the projected cost to complete the MOX facility exceeds $200 million. NNSA began withholding incentive fees for the first quarter of fiscal year 2011 when, for the first time, the increase in the projected cost to complete the facility exceeded this threshold. NNSA memos for subsequent quarters in fiscal year 2011 noted that the project's cost and schedule metrics continued to worsen, reducing the likelihood of resumption of payments. In a July 2011 letter to the contractor explaining its rationale for not resuming payments, NNSA stated that it was sensitive to the potential impacts of the "nuclear renaissance"—the contractor's term for the resurgence of U.S. nuclear engineering and manufacturing capability after being dormant for more than 20 years, which the contractor stated limited the availability of qualified suppliers and subcontractors and led to staff turnover and higher-than-anticipated costs to install equipment. However, NNSA stated that such impacts would not necessarily overcome other evidence showing that the contractor was not meeting the overarching goal of the incentive fees, which is that the facility be completed within cost.[20]

- *Award fees*. NNSA withheld $8.6 million or about a quarter of the $32.6 million in award fees that the MOX contractor could earn from fiscal year 2008 through fiscal year 2012. The amount withheld included about half of the fees the contractor was eligible to earn in fiscal year 2012. NNSA's award fee evaluation for fiscal year 2012 cited various factors, such as poor construction planning; less than optimal coordination of work; and overly conservative specifications for installation of fire doors, resulting in delays and unnecessary costs. In contrast, NNSA paid $24.0 million in award fees for performance in other areas, such as maintaining a high level of worker safety—an area in which the contractor has consistently performed well, according to NNSA's award fee evaluations.
- *Milestone fees*. NNSA did not withhold any milestone fees and instead paid milestone fees of $30.8 million for tasks with deadlines ranging from February 2009 to March 2014. Examples of tasks for which NNSA paid milestone fees (some of which the MOX contractor completed early) included completing the roof, installing the first glovebox, constructing a technical support building, and completing a

start-up plan for the facility. According to the NNSA officials, although NNSA did not withhold milestone fees, NNSA stopped paying any of the $30.2 million in remaining milestones fees as part of an understanding with the contractor to renegotiate the amount of and conditions for earning milestone fees.

- *Fixed fees*. According to the contracting officer, NNSA did not withhold any of the $15.7 million in fixed fees—the total amount of fixed fees for construction-related work under the MOX contract. NNSA included these fees in the contract to reward the contractor for work performed during contract negotiations, when other fees had not yet been negotiated.

NNSA Withheld 41 Percent of the WSB Contractor's Fees through Fiscal Year 2012

In a March 2013 analysis of the WSB contractor's performance, the NNSA contracting officer for the WSB recommended that the contractor should be held accountable for performance failures that contributed to the project's cost increase. For example, the analysis stated that the contractor did not require the subcontractor to add crews or take other steps to correct delays until almost 2 years after the federal project director began expressing concerns about the delays. In accordance with this assessment, NNSA withheld $7.7 million or about 40 percent of the $18.9 million in performance incentive fees that the WSB contractor could earn from fiscal year 2009, when construction began, through fiscal year 2012, for the portion of fees allocated to construction of the WSB under the management and operation contract for the Savannah River Site. Most of the fees withheld were for the contractor's performance in fiscal years 2011 and 2012 (see table 2). In particular, NNSA withheld $3.3 million of the $6.9 million in fees the contractor could earn in fiscal year 2011 and $3.9 million of the $4.0 million in fees the contractor could earn in fiscal year 2012. The fees withheld were tied to various performance measures, which DOE acquisition regulations require be established prior to the start of each evaluation period. Performance measures NNSA established included meeting the schedule for testing various types of equipment, providing engineering support to and coordinating with the construction subcontractor, and maintaining the project within preestablished cost and schedule metrics. The $3.3 million in fees withheld for fiscal year 2011 included $2 million that NNSA took back—that is, was paid back by the contractor—after making its fee determination for the contractor. Specifically, according to a December 2012 letter from the NNSA contracting officer to the contractor, the fiscal year 2011 fee determination was premised on the contractor's statements that schedule delays were recoverable and that the project would be completed within the approved cost estimate. Shortly after NNSA made its fee determination, however, the contractor notified NNSA that the project was further behind schedule than previously represented and that cost factors not included in the contractor's system for tracking and reporting on cost and schedule performance would result in a cost overrun. The contracting officer's letter stated that NNSA would have reduced the contractor's fee if it had known the extent of delays and cost overruns when it made its fee determination, and NNSA required the contractor to repay $4 million. In May 2013, NNSA agreed to a settlement with the contractor to reduce the amount taken back to $2 million after the contractor appealed NNSA's initial demand.

Table 2. Fees Paid and Withheld for Construction of the WSB through Fiscal Year 2012

Dollars in millions				
Fiscal year	**Amount paid**	**Amount withheld**	**Percentage withheld**	**Total available to be earned**
2009	$1.5	< $0.1[a]	2%	$1.6
2010	5.8	0.6	9	6.4
2011	3.7	3.3[b]	47	6.9
2012	0.2	3.9	96	4.0
Total[c]	**$11.2**	**$7.7**	**41%**	**$18.9**

Source: GAO analysis of NNSA data on fees paid to the WSB contractor.

Note: Dollar figures may not add up to totals due to rounding. [a]NNSA withheld $25,000 for fiscal year 2009.

[b]The fees withheld for fiscal year 2011 included $2 million that NNSA took back from the amount it originally paid to the WSB contractor.

[c]To provide an incentive to the contractor to improve its performance on the WSB, NNSA also set aside a total of $6 million in fees for the project in fiscal years 2013 and 2014. According to the NNSA contracting officer for the WSB, NNSA has not yet decided how to allocate fees for fiscal year 2015 under the overall management and operation contract for the Savannah River Site, including for the WSB project.

In addition to withholding fees, in a June 2012 letter to the contractor, NNSA's contracting officer questioned why she should not conclude that the contractor's actions rose to the level of gross negligence or willful misconduct, warranting disallowance of costs, meaning that the contractor would bear part of the cost increase resulting from the project's schedule delays.[21] For example, the letter stated that the contractor's system for tracking and reporting on cost and schedule performance did not meet industry standards and impeded NNSA's ability to understand the potential impact of delays in construction of various segments of the project on the final delivery date. According to NNSA officials, NNSA is waiting until after completion of WSB construction, and total construction costs are known, to determine unallowable costs.

ESTIMATES FOR THE PLUTONIUM DISPOSITION PROGRAM'S LIFE-CYCLE COST AND CONSTRUCTION PROJECTS DID NOT MEET ALL BEST PRACTICES FOR RELIABILITY

NNSA's most recent cost and schedule estimates for the Plutonium Disposition program did not fully reflect the characteristics of high-quality, reliable estimates as established by best practices used throughout government and industry and documented in the *GAO Cost Estimating and Assessment Guide* and *GAO Schedule Assessment Guide*. Specifically, (1) NNSA's draft April 2013 life-cycle cost estimate for the overall program was partially comprehensive, partially well-documented, and partially accurate but did not meet any of the best practices for a credible estimate; (2) the MOX contractor's September 2012 proposal for increasing the cost of the MOX facility was substantially comprehensive but partially well-documented and accurate and minimally credible; and (3) the WSB contractor's February 2013 monthly update to its schedule estimate was minimally well-constructed and partially met the other three characteristics of a reliable schedule—comprehensive, credible, and controlled.[22]

NNSA Did Not Follow All Key Steps for Developing a Reliable Life-cycle Cost Estimate for the Plutonium Disposition Program

In developing its draft April 2013 life-cycle cost estimate of $24.2 billion for the Plutonium Disposition program, NNSA followed several of the 12 key steps for developing high-quality cost estimates, including defining the estimate's purpose, defining the program's characteristics, and obtaining the data. NNSA did not follow other key steps, however, such as conducting an independent cost estimate. As a result, the estimate was not reliable. In particular, NNSA's draft life-cycle cost was partially comprehensive, partially well-documented, and partially accurate but did not meet any of the best practices for a credible estimate.

Table 3 summarizes the major components of NNSA's draft April 2013 life-cycle cost estimate. The estimate assumed that the MOX facility would start operations in November 2019 and that it would take approximately 15 years to complete the mission to dispose of 34 metric tons of surplus weapons-grade plutonium.

Table 3. NNSA's Draft Life-cycle Cost Estimate for the U.S. Plutonium Disposition Program, as of April 2013

Dollars in millions				
Facility or program component	**Type of cost**	**Actual costs[a]**	**Projected costs[b]**	**Total**
MOX Fuel	Construction	$3,435.6	$3,988.6	$7,424.2
Fabrication Facility	Operations and maintenance	2.7	8,256.1	8,258.8
WSB	Construction	265.1	132.7	397.9
	Operations and maintenance	-	1,910.2	1,910.2
PDCF	Construction	730.1	-	730.1
MIFT	Operations and maintenance	681.7	4,258.9	4,940.6
PDIP	Operations and maintenance	65.2	427.6	492.8
Total		**$5,180.6**	**$18,974.1**	**$24,154.7**

Source: GAO analysis of NNSA data.

Note: Dollar figures may not add up to totals due to rounding.

[a]Actual costs cover fiscal year 1999 through fiscal year 2012. Data is based on actual costs incurred.

[b]Projected costs cover fiscal year 2013 through fiscal year 2036. Data is based on projected funding required.

Table 4 lists the steps, or best practices, necessary for developing a high-quality cost estimate. Appendix II summarizes our assessment of NNSA's process for developing its draft life-cycle cost estimate against the steps that should result in the four characteristics of a high-quality cost estimate.

Table 4. Four Characteristics of a High-Quality Cost Estimate with Their Corresponding 12 Key Cost Estimating Steps (or Best Practices)

Characteristic	12 Steps (or best practices)
Comprehensive	• Develop the estimating plan • Determine the estimating structure
Well-documented	• Define the estimate's purpose • Define the program's characteristics • Identify ground rules and assumptions • Obtain the data • Document the estimate • Present the estimate to management for approval
Accurate	• Develop the point estimatea • Update the estimate to reflect actual costs and changes
Credible	• Compare the point estimate to an independent cost estimatea • Conduct a sensitivity analysis • Conduct a risk and uncertainty analysis

Source: GAO.

[a]As described in the *GAO Cost Estimating and Assessment Guide*, developing the point estimate and comparing it with an independent cost estimate are separate parts of a single step. For purposes of assessing the extent to which a cost estimate achieves the characteristics of a high-quality cost estimate, developing the point estimate contributes to accuracy, and comparing the point estimate with an independent cost estimate contributes to credibility.

Our assessment of NNSA's process for developing its draft life-cycle cost estimate included the following observations:

- *Comprehensive.* The draft life-cycle cost estimate was partially comprehensive because work breakdown structures were developed for the MOX and WSB projects and other components of the program, but NNSA had not formalized a program-level work breakdown structure. A typical work breakdown structure provides a clear picture of what needs to be accomplished, how the work will be done, and a basis for identifying resources and tasks for developing a cost estimate. Without a program-level work breakdown structure, NNSA cannot ensure that its life-cycle cost estimate captures all relevant costs, which can mean cost overruns.
- *Well-documented.* The draft life-cycle cost estimate was partially well-documented because NNSA defined the estimate's purpose and the program's characteristics, but it did not develop a single document

to describe data sources and steps taken in developing the estimate—such as applying escalation rates to account for inflation—so that the estimate could be replicated by someone other than those who prepared it. In addition, NNSA stated that a document identified the estimate's ground rules and assumptions but that the assumptions have changed frequently, hindering development of a life-cycle cost estimate. Examples of changes in assumptions not reflected in NNSA's draft April 2013 estimate included the slowdown of activities during the assessment of alternative plutonium disposition strategies and NNSA's plans to increase the amount of plutonium disposed of as MOX fuel.

- *Accurate.* The draft life-cycle cost estimate was partially accurate in that NNSA followed the best practice for developing a point estimate—a best guess at a cost estimate usually falling between best and worst case extremes. NNSA also updated the estimate periodically to include actual costs and changes to program and project requirements. However, NNSA did not use a formal system for tracking and reporting on cost and schedule performance to update the estimate, limiting the ability of someone other than those who prepared the estimate to check the estimate's accuracy and to identify when, how much, and why the program cost more or less than planned.
- *Credible.* The draft life-cycle cost estimate was not credible because NNSA did not conduct an independent cost estimate to provide an unbiased test of whether its estimate was reasonable, a formal sensitivity analysis to examine the effects of changing assumptions and ground rules, or a risk and uncertainty analysis to assess variability in point estimates due to factors such as errors and cost estimators' inexperience or biases. NNSA conducted such analyses for portions of its life-cycle cost estimate, but not for the entire estimate. For example, NNSA's Plutonium Disposition program office arranged for another office within NNSA to conduct an independent assessment of the MOX facility's operations costs, but not for the program's entire life-cycle cost.

NNSA did not follow all key steps for developing high-quality cost estimates in part because it did not have a requirement to develop its life-cycle cost estimate. According to NNSA officials, DOE's project management order includes requirements for development of cost and schedule estimates for a project, such as the MOX facility or WSB, but does not specify equivalent requirements for a program like Plutonium Disposition, which includes multiple projects, as well as supporting activities. As a result, when developing the life-cycle cost estimate for the Plutonium Disposition program, NNSA officials used an ad hoc approach to adapt requirements for managing projects in DOE's project management order. NNSA officials also said that its April 2013 life-cycle cost estimate did not include all the steps of a high-quality, reliable estimate in part because NNSA considered the estimate to be draft and, therefore, had not fully implemented plans for developing it.

In the absence of a specific requirement in DOE's project management order for developing a life-cycle cost estimate for a program, NNSA officials said they developed a life-cycle cost estimate for the Plutonium Disposition program for several reasons. According to these officials, these reasons included that the cost of the program is largely made up of capital projects, such as the MOX facility, and that requirements for congressional budget submissions specify that the full life-cycle cost of such projects be presented. In addition, each year NNSA must submit to Congress its estimated expenditures covering the fiscal year with respect to which the budget is submitted and at least the four succeeding fiscal years. NNSA officials said that, to accurately estimate expenditures for this 5-year period, they needed to develop a life-cycle cost estimate for the overall Plutonium Disposition program. An NNSA official noted that NNSA plans to use a version of its life-cycle cost estimate as a basis for evaluating alternative strategies to dispose of surplus weapons-grade plutonium.

Contractor's Proposed Estimate for the MOX Facility Did Not Meet Most Best Practices for Reliability

The MOX contractor's September 2012 proposal for increasing the cost of the MOX facility was substantially comprehensive but was partially well-documented, partially accurate, and minimally credible. The contractor's estimate did not fully reflect the characteristics of a high-quality, reliable estimate in part because it was a proposal, as opposed to an approved cost estimate. For example, one of the best practices for a well-documented estimate—and a requirement of DOE's project management order—is that a cost estimate be reviewed and accepted by management. Because DOE had not approved it and instead postponed its review and approval pending the outcome of NNSA's assessment of alternative plutonium disposition strategies, the contractor's estimate partially met this best practice. This best practice would be met by DOE's completion of its review and approval of a new estimate for the MOX facility, assuming the assessment of alternative plutonium disposition strategies maintains the current strategy of disposing plutonium as MOX fuel.

Though the contractor's September 2012 estimate did not fully reflect the characteristics of a high-quality estimate and cannot be considered reliable, the MOX contractor began using it as a provisional baseline for purposes of monthly reporting on the project's cost and schedule performance. Specifically, as directed by NNSA, the contractor began a transition in June 2012 to report its monthly performance against the contractor's proposed estimate of $7.7 billion. The contractor completed the transition and ceased any reporting of performance against the previously approved estimate early in 2013. Managing projects that no longer have an approved cost and schedule estimate is a challenge because cost and schedule estimates provide a baseline for measuring progress. At a July 2013 hearing, the Deputy Secretary of Energy noted that not having such a baseline is the point of maximum risk of unrestricted cost growth on a project.

Appendix III summarizes our assessment of how well the MOX contractor's proposal met the characteristics of a high-quality estimate. Our assessment included the following observations:

- *Comprehensive.* The proposal was substantially comprehensive in that it included all construction costs, as defined by the statement of work under the MOX contract. The proposal was not fully comprehensive, however, because it only partially met certain best practices for a comprehensive estimate, such as documenting all cost-influencing ground rules and assumptions. The proposal partially met this best practice because it did not provide justifications for some assumptions, such as not more than 10 percent of the supports for piping systems being nonstandard and requiring unique designs.
- *Well-documented.* The proposal was partially well-documented because it described in sufficient detail the calculations performed and the estimating methodology used to derive the cost of each element in the work breakdown structure. However, it did not provide all types of information specified in best practices for a well-documented estimate, such as how data on labor and travel costs were normalized. Data normalization is often necessary to ensure comparability because

data can be gathered from a variety of sources and in different forms that need to be adjusted before being used.

- *Accurate*. The proposal was partially accurate in that it appeared to adjust cost elements for inflation and contained only a few minor mistakes, but the contractor did not update its proposal with actual costs incurred after it developed the proposal and submitted it to NNSA in September 2012. NNSA and contractor officials agreed that the estimate was no longer an accurate reflection of the cost to complete construction—for example, because the proposal assumed a higher level of funding than the project received in fiscal year 2013. The officials said that, if the MOX project continues, the contractor would need to prepare a new proposal that includes costs for work conducted after the initial proposal was developed.
- *Credible*. The proposal was minimally credible because DOE halted its independent cost estimate of the proposal pending the outcome of NNSA's assessment of alternative plutonium disposition strategies. Moreover, the proposal did not include a formal sensitivity analysis to examine the effects of changing assumptions and ground rules, and it provided no evidence that major cost elements were cross-checked to determine whether alternative cost-estimating methods produced similar results. Finally, the proposal included an analysis of risks, such as difficulty attracting and retaining workers, and uncertainty in estimating materials and other costs. On the basis of this analysis, the proposal included a total of $713.1 million to account for risks and uncertainty—$641.4 million for the original scope of the MOX facility and $71.7 million for the addition of plutonium feedstock capability (see table 5). However, the contractor did not properly conduct or clearly document all steps in the analysis to determine the amount of funding to account for risks and uncertainty that could increase the cost of the project.

Contractor's Schedule Estimate for the WSB Did Not Meet Most Best Practices for Reliability

The WSB contractor's February 2013 monthly update to its schedule estimate did not fully reflect the characteristics of a high-quality, reliable schedule estimate as established by best practices. Specifically, the contractor's schedule estimate was minimally well-constructed and partially met the other three characteristics of a reliable, high-quality schedule as measured against best practices—comprehensive, credible, and controlled. Table 6 shows the characteristics of a high-quality schedule estimate and corresponding best practices. Appendix IV summarizes our assessment of how well the WSB contractor's February 2013 schedule estimate met the characteristics of a high-quality estimate.

Table 5. MOX Contractor's September 2012 Proposal for Increasing the Project's Cost

Dollars in millions		
Scope of work covered by the proposal	**Type of cost**	**Amount**
MOX Fuel Fabrication Facility	Construction cost and fees[a]	$6,770.9
	Allowances for risks and estimate uncertainty	641.4
Addition of plutonium feedstock capability	Construction cost and fees	190.6
	Allowances for risks and estimate uncertainty	71.7
Total		**$7,674.7**

Source: Shaw AREVA MOX Services, LLC.

Note: Dollar figures do not add up to $7,674.7 million due to rounding.

[a]The contractor's proposal included increases to incentive, milestone, and award fees specified under the MOX contract.

Table 6. Four Characteristics of a High-Quality Schedule Estimate with Their Corresponding Best Practices

Characteristic	Best practices
Comprehensive	• Capturing all activities • Assigning resources to all activities • Establishing the durations of all activities
Well-constructed	• Sequencing all activities • Confirming that the critical path is valid • Ensuring reasonable total float
Credible	• Verifying that the schedule is traceable horizontallyandvertically • Conducting a schedule risk analysis
Controlled	• Updating the schedule with actual progress and logic • Maintaining a baseline schedule

Source: GAO.

Our assessment of the WSB contractor's February 2013 schedule estimate included the following observations:

- *Comprehensive.* The estimate was partially comprehensive in that it captured and established the durations of contractor and government activities to complete the project but did not capture the remaining detailed work to be performed by the construction subcontractor. Specifically, it reduced the subcontractor's 3,851 activities to complete its portion of the work to one placeholder activity. According to the NNSA federal project director, the WSB contractor reduced the subcontractor's activities to a placeholder because the subcontractor submitted unreliable schedules with repeated changes in the estimated completion date for its portion of work.
- *Well-constructed.* The estimate was minimally well-constructed in that it sequenced activities in ways that can obscure a schedule's earliest completion date. In addition, the sequencing of activities included "merge points"—the convergence of many parallel activities into a single successor activity, which decreased the probability of successor activities starting on time. For example, performance of an assessment of readiness to operate the WSB was preceded by 212 activities. NNSA officials explained that the merge points resulted from the need to complete activities in parallel to meet requirements set forth in DOE's project management order.
- *Credible.* The estimate was partially credible in that the WSB contractor conducted a schedule risk analysis to determine the amount of schedule contingency—a reserve of extra time to account for risks and ensure completion of the project on time. However, a DOE review conducted prior to approval of an increase in the project's cost and a delay in the start of operations found that the results of the contractor's analysis were unreliable—for example, because project team members were not consulted regarding risk inputs. As a result, the schedule risk analysis did not clearly support the 12 months of schedule contingency included in the approved cost increase and schedule delay.
- *Controlled.* The estimate was partially controlled in that, according to project officials, the schedule was updated weekly and used to measure performance, but no narrative accompanied weekly updates to provide decision makers with a log of changes and their effect, if any, on the schedule time frame. In addition, project officials did not

provide documentation enabling the schedule to be validated, such as documentation describing sequencing of activities or assumptions used in developing the schedule.

The NNSA federal project director and contractor's project leader said that the contractor had begun to correct problems in the contractor's schedule estimate—for example, by replacing the placeholder for the subcontractor's activities with a schedule of more detailed activities independently developed by the contractor. However, delays on the project continued after the contractor began correcting the problems. Notably, according to DOE's October 2013 monthly report on the WSB, continuing delays in completion of the construction subcontract—one of the key drivers NNSA identified for the WSB cost increase—already used up about 10 of the 12 months of schedule contingency, placing the project's completion date in jeopardy.

CONCLUSION

NNSA has identified drivers of the close to $3 billion increase in the projected cost to complete the MOX facility and WSB and has taken steps to hold the MOX and WSB contractors accountable for their role in the cost increases by withholding and taking back fees. However, the various drivers identified by NNSA, such as DOE's approval of the cost and schedule estimate for the MOX facility before design was complete, do not provide the level of detail that can be found in a root cause analysis. In addition, DOE's project management order requires that lessons learned be captured throughout a project but does not include a requirement for a root cause analysis when a project exceeds its cost estimate, even when a project exceeds its cost estimate by billions of dollars. The decision whether to conduct such an analysis is instead made on a case-by-case basis. Because NNSA has not conducted a root cause analysis to identify the underlying causes of the cost increases for the MOX facility and WSB, it cannot provide assurance that it has correctly identified the underlying causes to ensure that they will not lead to further cost increases as the projects move forward. Further, without a root cause analysis, NNSA's ability to identify recommended solutions and lessons learned that could be applied to other projects is lessened. Conducting a root cause analysis of the cost increases for the MOX facility and WSB could help NNSA address its long-standing difficulties in completing projects within cost and on schedule, which has led to NNSA's project management remaining on GAO's list of areas at high risk of fraud, waste, abuse, and mismanagement.

NNSA has drafted a life-cycle cost estimate of $24.2 billion for the Plutonium Disposition program—an important step toward presenting the full cost of NNSA's current strategy to dispose of surplus weapons-grade plutonium as MOX fuel. A cost estimate that presents the full cost of NNSA's current plutonium disposition strategy is essential to inform NNSA's ongoing evaluation of alternative plutonium disposition strategies and provide Congress with a complete picture of the cost of the program. NNSA developed its life-cycle cost estimate even though neither DOE nor NNSA required the estimate. In particular, DOE's project management order does not explicitly require that life-cycle cost estimates be developed for programs like the Plutonium Disposition program that include both construction projects and other efforts and activities not related to construction, such as producing plutonium feedstock for the MOX facility. In the absence of such a requirement, NNSA followed several of the 12 key steps described in the *GAO Cost Estimating and Assessment Guide* for developing high-quality, reliable cost estimates, but it did not follow other key steps. Because NNSA did not follow all of the steps, the life-cycle estimate for the Plutonium Disposition program is not reliable. Similarly, the contractors' cost and schedule estimates for the MOX facility and WSB did not meet all best practices compiled in GAO's guides for preparing high-quality, reliable cost and schedule estimates. Not meeting these best practices increased the risk of further cost increases and delays for the projects and, because the projects are components of NNSA's life-cycle cost estimate, for the overall Plutonium Disposition program.

RECOMMENDATIONS FOR EXECUTIVE ACTION

We are making six recommendations in this report to the Secretary of Energy. To identify lessons learned from and provide assurance of preventing recurrence of cost increases for the MOX facility and WSB, and to develop reliable cost estimates for the Plutonium Disposition program, we recommend that the Secretary of Energy direct the DOE and NNSA Offices of Acquisition and Project Management and the NNSA office responsible for managing the Plutonium Disposition program, as appropriate, to take the following four actions:

- Conduct an analysis of the root causes of the cost increases for the MOX facility and WSB, such as the causes of the design changes that led to cost increases, and identify and prioritize recommended solutions.
- Revise and update the program's life-cycle cost estimate following the 12 key steps described in the *GAO Cost Estimating and Assessment Guide* for developing high-quality cost estimates, such as conducting an independent cost estimate to provide an objective and unbiased assessment of whether the estimate can be achieved.
- Ensure that the MOX contractor revises its proposal for increasing the cost of the MOX facility to meet all best practices for a high-quality, reliable cost estimate—for example, by cross-checking major cost elements to determine whether alternative estimating methods produce similar results.
- Ensure that the approved cost increase for the WSB is based on a schedule that the contractor has revised to meet all best practices for a high-quality, reliable schedule estimate, such as reflecting all activities (both government and contractor) needed to complete construction.

To ensure that future DOE projects benefit from lessons learned that reflect the underlying causes of cost increases or schedule delays experienced by other projects, and that Congress and DOE have life-cycle cost estimates for DOE programs that include individual construction projects, we further recommend that the Secretary of Energy take the following two actions to revise DOE's project management order or otherwise implement a departmentwide requirement:

- Require a root cause analysis of all projects that experience cost increases or schedule delays exceeding a certain threshold established by DOE.
- Require life-cycle cost estimates covering the full cost of programs that include both construction projects and other efforts and activities not related to construction.

AGENCY COMMENTS AND OUR EVALUATION

We provided a draft of this product to DOE for comment. In written comments, NNSA stated that the agency and DOE generally agreed with our recommendations. In particular, NNSA concurred with four of our six recommendations and partially concurred with the other two. NNSA described actions it planned to take to implement the recommendations with which it concurred and time frames for taking these actions. NNSA also provided technical comments, which we incorporated into the report as appropriate.

We are pleased that NNSA concurred with our recommendation to conduct an analysis of the root causes of the cost increases for the MOX facility and WSB and stated that it is planning to conduct such an analysis, which was not mentioned during the course of our review. NNSA also concurred with our recommendation to revise and update the Plutonium Disposition program's life-cycle cost estimate and stated that it would do so after a decision was made on the path forward for the program. The path forward could involve the use of alternative strategies to dispose of surplus weapons-grade plutonium.

NNSA also concurred with our recommendation to ensure that the MOX contractor revises its proposal for increasing the cost of the MOX facility to meet all best practices for a high-quality, reliable cost estimate. In its comment letter, NNSA stated that it is working with the contractor to ensure that the cost estimating processes and procedures are updated such that the best practices are met.

In addition, NNSA concurred with our recommendation to ensure that the approved cost increase for the WSB is based on a schedule that the contractor has revised to meet all best practices for a high-quality, reliable schedule estimate. In its comment letter, NNSA stated that it has revised the schedule since we reviewed it and that it now reflects all activities needed to complete construction. We did not review the update to the WSB contractor's schedule to confirm that it captured all activities to complete construction, which is one of the best practices associated with the characteristics of a high-quality schedule. Moreover, as detailed in appendix IV, the schedule we reviewed only partially or minimally met 7 of the other 9 best practices. To fully implement our recommendation, NNSA would need to ensure that the contractor has revised its schedule to meet all best practices for a high-quality, reliable schedule estimate. In its comment letter, NNSA stated that during the next project review, which is expected to occur by December 31, 2014, NNSA will review the schedule against best practices.

NNSA partially concurred with our fifth recommendation that DOE require a root cause analysis of all projects that experience cost increases or schedule delays exceeding a certain threshold established by the department. In its comment letter, NNSA stated that DOE program offices currently perform tailored root cause analyses as part of the baseline change proposal process outlined in the department's project management order for increasing a project's cost and schedule estimates. NNSA stated that, as a result, the department does not believe that an update to the project management order is required. NNSA further stated that the department will review the lessons learned from NNSA's root cause analyses for the MOX and WSB projects to see what best practices may be of benefit to other projects. However, as we stated in the report, DOE's project management order does not include a requirement for a root cause analysis of projects experiencing significant cost increases or schedule delays, and NNSA officials said that they decide on a case-by-case basis whether to conduct a root cause analysis. Moreover, the order does not define what a root cause analysis is, how or when a root cause analysis should be conducted, or what is meant by a tailored analysis. In addition, NNSA's written comments did not provide information on the conditions that would trigger a root cause analysis. Leaving root cause analyses to an informal and undefined process within DOE program offices could result in such analyses not being conducted, not being conducted consistently, or not accurately identifying underlying causes of cost increases in order to identify and implement corrective measures and apply lessons learned to other DOE projects. We continue to believe that a root cause analysis should be conducted for all projects that experience cost increases or schedule delays above a threshold established by the department. We note that our recommendation is consistent with a requirement in the Weapon Systems Acquisition Reform Act of 2009; under the act, the Department of Defense must perform a root cause analysis of a cost increase that exceeds a certain threshold.

NNSA partially concurred with our final recommendation that DOE require life-cycle cost estimates covering the full cost of programs that include both construction projects and other efforts and activities not related to construction. In its comment letter, NNSA stated that the department's project management order requires a comprehensive life-cycle cost analysis as part of the alternative selection process and that no further update to the order is required to address this recommendation. The intent of our recommendation goes beyond that of preparing a life-cycle cost estimate at the stage of selecting an alternative for a new capital asset project. Instead, the recommendation applies to departmental programs that include capital asset projects to meet the overall program need. As we stated in the report, NNSA did not follow all key steps for developing high-quality cost estimates in developing its draft April 2013 life-cycle cost estimate for the Plutonium Disposition program, which currently includes the MOX facility and WSB capital asset projects, in part because it did not have a requirement to develop it. NNSA's response to our recommendation suggests that the life-cycle cost estimates for the MOX and WSB projects that were required to be prepared years ago, when the projects were selected from among other alternatives, are the only life-cycle cost estimates needed to manage the Plutonium Disposition program. Furthermore, NNSA's response contradicts the fact that it concurred with our recommendation to revise and update the life-cycle cost estimate for the overall Plutonium Disposition program in accordance with cost estimating best practices. We continue to believe that our recommendation that the department require life-cycle cost estimates covering the full cost of programs that include construction projects should be implemented.

David C. Trimble
Director, Natural Resources and Environment

APPENDIX I: OBJECTIVES, SCOPE, AND METHODOLOGY

To assess drivers of the construction cost increases for the Mixed Oxide (MOX) Fuel Fabrication Facility and Waste Solidification Building (WSB) that the National Nuclear Security Administration (NNSA) identified, we reviewed the Department of Energy's (DOE) budget request for NNSA for fiscal year 2014, which provided a summary of the cost drivers for both projects.

To assess cost drivers in further detail, we reviewed the MOX contractor's September 2012 proposal for increasing the project's cost, which discussed drivers from the contractor's perspective. We also reviewed DOE's December 2012 document approving an increase in the estimated cost of the WSB and a delay in the start of operations, which summarized cost drivers and their impact on the project's cost and schedule.

We visited the Savannah River Site to observe construction progress for both projects and interviewed NNSA and contractor officials responsible for managing the projects. We also interviewed officials from the NNSA Office of Fissile Materials Disposition, the NNSA Office of Acquisition and Project

Management, and the DOE Office of Acquisition and Project Management. Separately, to understand how, if at all, cost drivers for the MOX facility were related to Nuclear Regulatory Commission (NRC) regulation and licensing of the construction and operation of the facility, we reviewed NRC construction inspection reports and related documents, and we interviewed NRC officials responsible for overseeing the facility's construction.

In order to understand the components of cost growth for the MOX facility, which represented most of the Plutonium Disposition program's construction cost increase, we also analyzed the MOX contractor's earned value management (EVM) system that the contractor used to track and report on cost and schedule performance, including data from the EVM system on cumulative cost and schedule variance trends from July 2011 through April 2012 and the contractor's variance report for April 2012.

To determine the extent to which NNSA analyzed underlying causes of the cost increases, we reviewed documents providing context for cost drivers. The documents we reviewed included NNSA Office of Acquisition and Project Management project review reports and monthly status reports; DOE Office of Acquisition and Project Management monthly status reports; DOE documents related to approval of the previous cost and schedule estimates for the MOX facility and WSB in April 2007 and December 2008, respectively; and documents related to specific cost drivers identified by NNSA, such as the MOX contractor's October 2006 report on construction markets and DOE reports related to its suspension of the WSB contractor's system for tracking and reporting cost and schedule performance in November 2012. We also interviewed NNSA officials to determine the extent to which they had conducted or planned any analyses to identify underlying causes of cost increases for the Plutonium Disposition program's construction projects.

To determine steps NNSA took to hold contractors accountable for their role in the cost increases for the Plutonium Disposition program's construction projects, we reviewed the contracts for the MOX facility and WSB, fees specified under the contracts, and NNSA's fee evaluations and other documentation supporting its fee determinations.

We also interviewed NNSA contracting officers who were responsible for administering the MOX and WSB contracts regarding the terms of the contracts, fees specified under the contracts, and actions NNSA took or planned to take to hold contractors accountable for their role in the cost increases.

We obtained NNSA data on fees it paid to and withheld from the contractors, and we assessed the reliability of the data by checking for obvious errors in accuracy and completeness; comparing the data with other sources of information, such as NNSA's fee determinations; and interviewing NNSA contracting officers who had knowledge of the data. We determined that NNSA's data on fees were sufficiently reliable for reporting on the fees paid to and withheld from the contractors.

To assess the extent to which NNSA's most recent estimates of the Plutonium Disposition program's life-cycle cost and of the cost and schedule for completing the program's construction projects met best practices we have compiled in guides identifying the characteristics of high-quality, reliable cost and schedule estimates,[1] we tailored our methodology to the differing stages of NNSA's development and approval of each estimate:

- *NNSA's life-cycle cost estimate for the Plutonium Disposition program.* Because NNSA had not finalized a life-cycle cost estimate, we assessed NNSA's most recent available estimate—spreadsheets dated April 2013 representing NNSA's draft life-cycle cost estimate. In particular, we assessed the process NNSA used to develop the estimate against the 12 key steps described in the *GAO Cost Estimating and Assessment Guide* that should result in a high-quality, reliable cost estimate. To provide information on NNSA's process, NNSA officials responsible for developing the estimate filled out a data collection instrument we developed. The data collection instrument summarized each of the 12 key steps and provided space for NNSA officials to describe actions they had taken to meet the criteria for each step. To review the information provided by NNSA, we checked NNSA's April 2013 estimate for obvious errors in accuracy and completeness and compared it with previous versions of the life-cycle cost estimate provided by NNSA.
  In addition, we interviewed NNSA officials to determine what requirements, if any, they followed for developing the estimate, their purpose for developing it, and their plans for presenting it for management approval. Finally, we interviewed NNSA officials from the Office of Analysis and Evaluation, which the Plutonium Disposition program had tasked with conducting an independent assessment of the MOX facility's operating costs.
- *NNSA's estimate to complete the MOX facility.* Because NNSA had not approved a revised cost and schedule estimate for the MOX facility, we assessed the MOX contractor's September 2012 proposal for increasing the project's cost, which NNSA had directed the MOX

contractor to use as a provisional baseline for purposes of monthly reporting. We compared data presented in various tables of the proposal for consistency and reviewed additional documents, including the technical baseline providing a detailed description of the MOX facility. We provided a draft of our assessment to NNSA and revised the draft, as appropriate, after discussing our assessment with NNSA program officials and the contractor.

- *NNSA's estimate to complete the WSB*. We assessed the WSB schedule estimate that the cost increase for the project approved in December 2012 was based on because, as described in the *GAO Schedule Assessment Guide*, a reliable schedule can contribute to an understanding of the cost impact if a project does not finish on time. Specifically, we compared the contractor's February 2013 monthly update to its schedule estimate, which was the most recent available update when we conducted our analysis, with the 10 best practices associated with the characteristics of a high-quality schedule. As part of our assessment, we reviewed documents related to the project's schedule, including NNSA's project execution plan for the WSB, the project's work breakdown structure, and the project's February 2013 update to the document showing the longest path to project completion.

  In addition, we interviewed the NNSA federal project director for the WSB and the WSB contractor's project leader and scheduler. We provided a draft of our assessment to NNSA and revised the draft, as appropriate, after discussing our assessment with NNSA program officials and the contractor.

We conducted this performance audit from November 2012 to February 2014 in accordance with generally accepted government auditing standards. Those standards require that we plan and perform the audit to obtain sufficient, appropriate evidence to provide a reasonable basis for our findings and conclusions based on our audit objectives.

We believe that the evidence obtained provides a reasonable basis for our findings and conclusions based on our audit objectives.

APPENDIX II: ASSESSMENT OF NNSA'S PROCESS FOR DEVELOPING A LIFE-CYCLE COST ESTIMATE FOR THE PLUTONIUM DISPOSITION PROGRAM

Best practice characteristic and overall assessment	Key steps to develop a high-quality cost estimate	Detailed assessmenta
Comprehensive: *Partially met*	Develop the estimating plan	*Partially met.* NNSA assigned a team to develop and update the estimate but did not have a written plan for developing it.
	Determine the estimating structure	*Partially met.* Work breakdown structures to define in detail the work necessary to accomplish objectives were developed for the MOX and WSB projects and other components of the program, but NNSA had not formalized a program-level work breakdown structure.
Well-documented: *Partially met*	Define the estimate's purpose	*Substantially met.* NNSA officials described the purpose as supporting annual budget requests, which include requirements that NNSA (1) present the full life-cycle cost of capital projects, such as the MOX facility, and (2) estimate expenditures for the fiscal year with respect to which the budget is submitted and at least the four succeeding fiscal years.
	Define the program's characteristics	*Substantially met.* NNSA developed a program requirements document to identify the scope, functions, and requirements of the Plutonium Disposition program. NNSA documented performance characteristics for program components in contracts, technical baselines, and execution plans.

Appendix II. (Continued)

Best practice characteristic and overall assessment	Key steps to develop a high-quality cost estimate	Detailed assessmenta
	Identify ground rules and assumptions	*Partially met.* NNSA identified ground rules and assumptions, but NNSA officials said that assumptions for the program change frequently, hindering development of a life-cycle cost estimate.
	Obtain the data	*Substantially met.* NNSA collected data at the project level, where,according to NNSA, data were documented in contractor systems and estimates were developed by teams of knowledgeable staff using historical information, current cost and pricing information, engineering and vendor quotes, cost guides, and current material and labor costs.
	Document the estimate	*Minimally met.* NNSA documented the estimate on spreadsheets, but it did not develop a single document to describe data sources and steps taken in developing the estimate so that it could be replicated by someone other than those who prepared it.
	Present the estimate to management for approval	*Not met.* NNSA considered the estimate to be draft and predecisional, and NNSA officials said they did not have plans to present an estimate to management for approval until NNSA completes its reevaluation of its strategy for disposing of surplus weapons-grade plutonium.
Accurate: *Partially met*	Develop the point estimateb	*Partially met.* NNSA developed a point estimate, but it did not use a program-level work breakdown structure to do so because it hadnot formalized such a structure.

Best practice characteristic and overall assessment	Key steps to develop a high-quality cost estimate	Detailed assessmenta
	Update the estimate to reflect actual costs and changes	*Partially met.* NNSA updated the estimate periodically to include actual costs and changes to program and project requirements, but it did not clearly document how changes affected the estimate.
Credible: *Not met*	Compare the point estimate to an independent cost estimateb	*Not met.* NNSA did not conduct an independent cost estimate for the overall program's life-cycle cost estimate, and it had not completed independent cost estimates for the program's two construction projects.
	Conduct a sensitivity analysis	*Not met.* NNSA did not conduct a formal sensitivity analysis at theprogram level.
	Conduct a risk and uncertainty analysis	*Not met.* NNSA did not conduct a risk and uncertainty analysis at the program level.

Source: GAO analysis of information from NNSA.

[a]The ratings we used in this analysis are as follows: "Not met" means NNSA provided no evidence that satisfies the key step.

"Minimally met" means NNSA provided evidence that satisfies a small portion of the key step. "Partially met" means NNSA provided evidence that satisfies about half of the key step. "Substantially met" means NNSA provided evidence that satisfies a large portion of the key step. "Fully met" means NNSA provided complete evidence that satisfies the entire key step.

[b] As described in the *GAO Cost Estimating and Assessment Guide*, developing the point estimate and comparing it with an independent cost estimate are separate parts of a single step.

For purposes of assessing the extent to which a cost estimate achieves the characteristics of a high-quality cost estimate, developing the point estimate contributes to accuracy, and comparing the point estimate with an independent cost estimate contributes to credibility.

Appendix III: Assessment of the MOX Contractor's Proposed Cost Estimate Compared with Industry Best Practices

Best practice characteristic and overall assessment	Best practice	Detailed assessmenta
Comprehensive: *Substantially met*	The cost estimate includes all life-cycle costs.	*Met.* The estimate covered construction and startup costs; at NNSA's direction, the estimate excluded operation and maintenance costs.
	The cost estimate completely defines the program, reflects the current schedule, and is technically reasonable.	*Substantially met.* The estimate was based on NNSA's statement of work and the contractor's technical baseline for the original scope of the MOX facility.
	The cost estimate work breakdown structure is product-oriented, traceable to the statement of work/objective, and at an appropriate level of detail to ensure that cost elements are neither omitted nor double-counted.	*Partially met.* The work breakdown structure clearly outlined the end product and major work of the project, but some cost elements were missing from the work breakdown structure.
	The estimate documents all cost-influencing ground rules and assumptions.	*Partially met.* The estimate documented that it was based on a profile of NNSA's projected annual funding to complete the project but did not provide justificationsfor some assumptions, such as not more than a set amount of work being nonstandard.

Best practice characteristic and overall assessment	Best practice	*Detailed assessmenta*
Well-documented: *Partially met*	The documentation captures the source data used, the reliability of the data, and how the data were normalized.	*Partially met.* The estimate was based on actual costs through May 2012 and used a database of labor and other costs, but it did not state whether or how all data had been normalized to ensure data comparability.
	The documentation describes in sufficient detail the calculations performed and the estimating methodology used to derive each element's cost.	*Met.* The estimate used a combination of expert opinion and extrapolation from actual data to develop estimates for and sum up individual cost elements of the work breakdown structure.
	The documentation describes, step by step, how the estimate was developed so that a cost analyst unfamiliar with the program could understand what was done and replicate it.	*Partially met.* The estimate used quantities of materialsand labor hours to develop estimates for individual cost elements but did not document how these quantities were estimated.
	The documentation discusses the technical baseline description and the data in the baseline is consistent with the estimate.	*Partially met.* The estimate agreed with NNSA's statement of work and the contractor's technical baseline for the original scope of the MOX facility, but the technical baseline did not cover the addition of capability to supply plutonium feedstock.
	The documentation provides evidence that the cost estimate was reviewed and accepted by management.	*Partially met.* DOE began a review of the proposed estimate but did not approve it.

Appendix III. (Continued)

Best practice characteristic and overall assessment	Best practice	Detailed assessmenta
Accurate: *Partially met*	The cost estimate results are unbiased, not overly conservative or optimistic, and based on an assessment of most likely costs.	*Minimally met.* The estimate was higher than needed to achieve an 85 percent confidence level—the level directed by NNSA—that the final cost would be less than the estimate.
	The estimate has been adjusted properly for inflation.	*Substantially met.* The estimate appeared to adjust cost elements for inflation, but adjustments were not well-documented.
	The estimate contains few, if any, minor mistakes.	*Met.* The estimate contained few minor mistakes, and calculations within the estimate were internally consistent.
	The cost estimate is regularly updated to reflect significant changes in the program so that it always reflects current status.	*Partially met.* The estimate was based on actual costs through May 2012 and did not reflect updated costs from the contractor's system for tracking and reporting cost and schedule performance.
	Variances between planned and actual costs are documented, explained, and reviewed.	*Minimally met.* The estimate explained variances between planned and actual costs at a high level but not at the cost element level.
	The estimate is based on a historical record of cost estimating and actual experiences from other comparable programs.	*Partially met.* The estimate did not explain to what extent it was based on historical data from other similar programs or facilities.

Best practice characteristic and overall assessment	Best practice	Detailed assessmenta
	The estimating technique for each cost element was used appropriately.	*Substantially met.* The estimating method used—developing the estimate at the lowest level of the work breakdown structure, one piece at a time, with the sum of the pieces becoming the estimate—was appropriate for a project under way.
Credible: *Minimally met*	The cost estimate includes a sensitivity analysis that identifies a range of possible costs based on varying major assumptions, parameters, and data inputs.	*Not met.* The estimate did not include a sensitivity analysis.
	A risk and uncertainty analysis was conducted that quantified the imperfectly understood risks and identified the effects of changing key cost driver assumptions and factors.	*Partially met.* The estimate included a risk and uncertainty analysis but did not properly conduct or clearly document all steps in the analysis.
	Major cost elements were cross-checked to see whether results were similar.	*Not met.* The estimate provided no evidence that majorcost elements were cross-checked.
	An independent cost estimate was conducted by a group outside the acquiring organization to determine whether other estimating methods produce similar results.	*Not met.* DOE halted its independent cost estimate of the contractor's proposed estimate as part of DOE's decision to reevaluate its strategy for disposing of surplus weapons-grade plutonium.

Source: GAO analysis of the MOX contractor's September 2012 proposal for increasing the cost of the MOX facility.

[a]The ratings we used in this analysis are as follows: "Not met" means the proposal provided no evidence that satisfies the best practice. "Minimally met" means the proposal provided evidence that satisfies a small portion of the best practice.

"Partially met" means the proposal provided evidence that satisfies about half of the best practice. "Substantially met" means the proposal provided evidence that satisfies a large portion of the best practice. "Fully met" means the proposal provided complete evidence that satisfies the entire best practice.

APPENDIX IV: ASSESSMENT OF THE WASTE SOLIDIFICATION BUILDING'S SCHEDULE ESTIMATE COMPARED WITH INDUSTRY BEST PRACTICES

Best practice characteristic and overall assessment	Best practice	Detailed assessmenta
Comprehensive: *Partially met*	Capturing all activities	*Minimally met.* The schedule estimate's 2,429 activities to complete the project included one summary activity in place of the construction subcontractor's 3,851 activities and, therefore, did not capture the remaining detailed work to be performed by the subcontractor.
	Assigning resources to all activities	*Partially met.* The schedule estimate assigned resources, such as labor and materials, to only about half of the remaining 2,429activities.
	Establishing the durations of all activities	*Substantially met.* The schedule estimate included activity durations that were generally short enough to be consistent withthe needs of effective planning.
Well-constructed: *Minimally met*	Sequencing all activities	*Minimally met.* The schedule estimate sequenced activities in ways that decreased the probability of activities starting on time and contained activities that were not properly tied with the start or end date of other activities, potentially obscuring the critical path determining the project's earliest completion date.

Best practice characteristic and overall assessment	Best practice	Detailed assessmenta
	Confirming that the critical path is valid	*Partially met.* Changes to the critical path were evaluated monthly and tracked in monthly status reports, but constraints inscheduled dates of certain activities convoluted the critical path.
	Ensuring reasonable total float	*Minimally met.* The schedule estimate included high total float values—the amount of time by which an activity can slip without affecting a completion date—potentially resulting in an inaccurate assessment of the project's completion date.
Credible: *Partially met*	Verifying that the schedule is traceable horizontally and vertically	*Substantially met.* The schedule estimate was traceable horizontally (i.e., across sequenced activities) and vertically (i.e., between activities and subactivities).
	Conducting a schedule risk analysis	*Minimally met.* The contractor conducted a schedule risk analysis, but the results of the analysis were unreliable for determining the likelihood of the project's completion date and did not align with DOE's revised cost and schedule estimate.
Controlled: *Partially met*	Updating the schedule with actual progress and logic	*Partially met.* According to project officials, the schedule was updated weekly, but no narrative accompanied the weekly updates.
	Maintaining a baseline schedule	*Minimally met.* Project officials stated that they used the schedule to measure performance, but they did not provide thorough documentation enabling the schedule to be validated, such as a narrative providing a log of changes and their effects.

Source: GAO analysis of the WSB contractor's schedule estimate, as updated in February 2013.

[a]The ratings we used in this analysis are as follows: "Not met" means the schedule estimate provided no evidence that satisfies the best practice. "Minimally met" means the schedule estimate provided evidence that satisfies a small portion of the best practice. "Partially met" means the schedule estimate provided evidence that satisfies about half of the best practice. "Substantially met" means the schedule estimate provided evidence that satisfies a large portion of the best practice. "Fully met" means the schedule estimate provided complete evidence that satisfies the entire best practice.

End Notes

[1] MOX fuel is a mix of plutonium and uranium oxides.

[2] The contractor for the MOX facility is Shaw AREVA MOX Services, LLC. The contractor for the WSB is Savannah River Nuclear Solutions, LLC. Design and construction management of the WSB is included in the contract between Savannah River Nuclear Solutions and DOE for management and operation of the Savannah River Site.

[3] DOE, *Program and Project Management for the Acquisition of Capital Assets*, DOE Order 413.3B (Washington, D.C.: Nov. 29, 2010).

[4] GAO, *High-Risk Series: An Update*, GAO-13-283 (Washington, D.C.: February 2013). In our 2013 High-Risk Update, we narrowed the focus of NNSA's high-risk designation to major contracts and projects, those with values of at least $750 million, to acknowledge progress made in managing smaller value efforts.

[5] GAO, *Department of Energy: Concerns with Major Construction Projects at the Office of Environmental Management and NNSA*, GAO-13-484T (Washington, D.C.: Mar. 20, 2013).

[6] In December 2008, as a result of funding reductions for fiscal year 2008, DOE approved a revised cost estimate for the MOX facility of $4.9 billion and a 1-month delay in the start of operations to October 2016.

[7] As described in NNSA's acquisition strategy, the WSB's start of operations included a buffer to allow for delays without having an impact on start-up testing of the MOX facility.

[8] GAO, *GAO Cost Estimating and Assessment Guide: Best Practices for Developing and Managing Capital Program Costs*, GAO-09-3SP (Washington, D.C.: March 2009); and GAO, *GAO Schedule Assessment Guide: Best Practices for Project Schedules*, GAO-12-120G (Washington, D.C.: May 2012).

[9] GAO, *Nuclear Nonproliferation: DOE Needs to Address Uncertainties with and Strengthen Independent Safety Oversight of Its Plutonium Disposition Program*, GAO-10-378 (Washington, D.C.: Mar. 26, 2010).

[10] DOE, *Draft Surplus Plutonium Disposition Supplemental Environmental Impact Statement*, DOE/EIS-0283-S2 (Washington, D.C.: July 2012).

[11] A cost is allowable if, among other things, it is reasonable, allocable to the contract at issue, and meets certain accounting standards and the terms of the contract. Federal Acquisition Regulation (FAR), 48 C.F.R. § 31.201-2 (2013). In general, a cost is allocable to a given contract if it is incurred specifically for the contract; benefits both the contract and other work, and can be distributed to them in reasonable proportion to the benefits received; or is necessary to the overall operation of the business, although a direct relationship to any particular cost objective cannot be shown. FAR, 48 C.F.R. § 31.201-4 (2013).

[12] GAO-09-3SP.

[13] GAO-12-120G.

[14] The NRC, an independent oversight agency, has general licensing and related regulatory authority over the construction and operation of the MOX facility. 42 U.S.C. § 5842 (2013). NRC issued a construction authorization for the facility in 2005. A license to operate the facility will only be issued by the NRC once it verifies that the facility's principal structures, systems, and components are properly constructed.

[15] Nuclear quality assurance criteria cover the design, fabrication, construction, and testing of structures, systems, and components used in nuclear power plants and other nuclear facilities. 10 C.F.R. pt. 50, app. B (2013).

[16] The subcontract included all construction activities for the WSB with the exception of early site work, such as installation of underground utilities.

[17] As described by DOE in an April 2008 report on its departmentwide effort to assess underlying causes for its project management challenges, a root cause analysis is a process involving the individuals knowledgeable of and directly responsible for managing DOE contracts and projects answering a challenging series of questions as to why a situation, event, or condition existed. The process continues with the identification, prioritization, and implementation of recommended solutions or corrective measures.

[18] For example, NNSA tasked a construction management contractor to identify the root causes associated with the need to increase the height of the Uranium Processing Facility planned for construction in Oak Ridge, Tennessee, to ensure that processing equipment would fit into the facility. The change to the facility's design resulted in approximately $540 million in additional costs. The root cause analysis focused not only on identifying lessons learned but also on understanding why the space within the facility became a major design issue late in the final design and recommending actions to fix the identified causes to prevent recurrence of similar negative outcomes. For further information on the cost increase for this facility, see GAO, *Nuclear Weapons: Factors Leading to Cost Increases with the Uranium Processing Facility*, GAO-13-686R (Washington, D.C.: July 12, 2013).

[19] The NNSA contracting officer for the MOX facility said that, as an alternative to taking back the provisional fees, NNSA may seek to use the $21.6 million to modify the MOX contract's cost-sharing provision to provide the contractor an incentive to limit further cost increases. The contract included a provision whereby the contractor would earn a portion of every dollar that total allowable, final project costs are less than the project's approved cost estimate, as adjusted for purposes of calculating the amount of cost share. In contrast, the contractor would owe a portion of every dollar that final project costs exceed the project's approved cost estimate, as adjusted.

[20] In technical comments on a draft of this report, NNSA stated that, because it hired a contractor with knowledge of both the marketplace for nuclear work and the MOX project requirements, it expected project estimates submitted by the contractor to account for any risk associated with the nuclear renaissance.

[21] In order for the government to reimburse a contractor's costs, the costs must be reasonable. FAR, 48 C.F.R. § 31.201-2, -3 (2013). Evidence of gross negligence or willful misconduct could provide a basis for a contracting officer or court to disallow costs as unreasonable under the regulation, but a contractor's actions do not need to rise to this level to be determined unreasonable, *See* Kellogg Brown & Root Servs. v. United States, 2013 U.S. App. LEXIS 18447, at *21-2, (D.C. Cir. Sept. 5, 2013).

[22] We based the ratings used in our analysis on the average of ratings for individual best practices included in each characteristic: "Not met" means an estimate provided no evidence that

satisfied the best practice. "Minimally met" means an estimate provided evidence that satisfied a small portion of the best practice. "Partially met" means the estimate provided evidence that satisfied about half of the best practice. "Substantially met" means the estimate provided evidence that satisfied a large portion of the best practice. "Fully met" means the estimate provided complete evidence that satisfied the entire best practice.

End Note for Appendix I

[1] GAO-09-3SP and GAO-12-120G.

INDEX